ENVIRONMENTAL ENRICHMENT AND REFINEMENT FOR NONHUMAN PRIMATES KEPT IN RESEARCH LABORATORIES:

A Photographic Documentation and Literature Review

(THIRD EDITION)

by Viktor and Annie Reinhardt

Animal Welfare Institute

ENVIRONMENTAL ENRICHMENT

is the provision of animate, inanimate and nutritional environmental modifications that promote the expression of species-appropriate behaviors (e.g., foraging) and species-appropriate mental activities (e.g., learning to cooperate during procedures) in relatively barren and boring living quarters.

REFINEMENT

is any modification in the housing and handling practices of animals that reduces or eliminates the subject's *distress* response to a specific condition (e.g., permanent single-housing) or situation (e.g., enforced restraint during a life-threatening procedure), and/or enhances the subject's well-being (e.g., promotes the expression of species-adequate behaviors in relatively barren living quarters).

DISTRESS

is the inability to adapt to a condition (e.g., barren cage) or to a situation (e.g., enforced restraint) that induces a conspicuous alteration in the subject's physiological equilibrium (e.g., significantly increased blood pressure) and/or psychological equilibrium (e.g., intense fear, self-directed aggression, hair pulling).

Animal Welfare Institute
P.O. Box 3650
Washington, DC 20027
www.awionline.org

Environmental Enrichment and Refinement for Nonhuman Primates Kept in Research Laboratories: A Photographic Documentation and Literature Review (Third Edition)

Printed in the United States of America

ISBN 978-0-938414-92-6
LCCN 2008907303

Cover photo by Derek Dammann
Design by Ava Rinehart
Copy editing by Cat Carroll and Cathy Liss

Mixed Sources
Product group from well-managed forests, controlled sources and recycled wood or fiber
FSC www.fsc.org Cert no. SW-COC-002504

TABLE OF CONTENTS

1. PREFACE

photo 1

Anyone who observes nonhuman primates in their natural habitat will quickly notice that the animals:

1. maintain constant vocal, optical or physical contact with other conspecifics;
2. show a distinct vertical flight response during alarming situations;
3. retreat to high places during the night; and
4. spend most of the day searching for, retrieving and processing food.

Seeing the inside of a primate research facility for the first time was a shocking experience for me, not only as a psychologically healthy person, but also as a scientist who has been trained to rigorously control extraneous variables that might influence research data. There were hundreds of animals kept in barren single-cages with nothing to do but stare at bleak walls and wait for their turn to be subjected with force to life-threatening procedures.

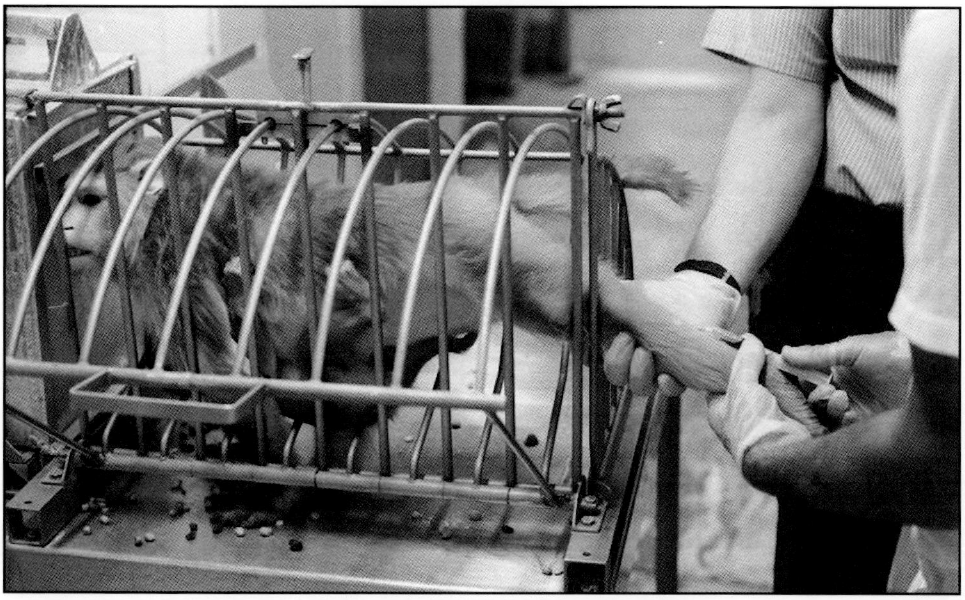

photo 2

The cages were stacked on top of each other in double-tiers to accommodate maximum numbers of animals in windowless rooms.

photo 3

The following poem, written by an animal technician of a prestigious primate research laboratory, puts exactly into words how I felt.

Hope Dashed

Walking, dazed
past cage and cage and cage
each contained an emotion
fear, depression and rage

each unique
one aggressive, the next is meek
a thousand lives locked away
with futures bleak

in stainless steel
a world surreal
no friend to touch
or sun to feel

entire lives kept complete
in 4.3 square feet
from birth through life
till last heartbeat.

It so happened that I soon got the opportunity to work in such a laboratory as clinical veterinarian and ethologist. I was determined to explore refinement strategies that would:

1. allow the animals to actively express their need for social contact and social interaction with at least one compatible conspecific (*Animate Enrichment*),
2. allow the animals to spend some time of the day searching for, retrieving and processing food (*Feeding Enrichment*),
3. open up the vertical dimension of the standard cage for the animals (*Structural Enhancement*),
4. give the animals a chance to cooperate rather than resist during procedures (*Positive Reinforcement Training*),
5. provide the animals with gadgets that distract them from boredom (*Inanimate Enrichment*).

In this third edition, the collection of photos is expanded and includes all nonhuman primates species.

The following individuals have kindly contributed photos for this book: James Anderson, Kate Baker, Ben Basile, Samantha Bjone, Hannah Buchanan-Smith, Moshe Bushmitz, Kathy Calligari, Arnold Chamove, Caroly Crockett, Natasha Down, Katherine Eckert, Jennifer Green, R. Goburdhun, Alison Kulick, Richard Lynch, Evan MacLean, Jean McKinley, Peggy O'Neill-Wagner, Kai Perret, Jillann Rawlins, Valerie Schoof, Jaylan Turkkan, Elaine Videan, Margaret Whittaker, and Andrew Winterborn.

I am grateful to Cat Carroll and Cathy Liss for pointing out errors in the manuscript and correcting grammatical flaws.

For the review of the literature, I have included only material that is of practical relevance, and/or is supported by animal welfare-relevant scientific data.

It is my wish to inspire animal care personnel, veterinarians and biomedical investigators to allow themselves to feel compassion for the animals with whom they work and to have the courage to translate these feelings into action.

Viktor Reinhardt
Mt. Shasta, California
May 2008

2. ANIMAL WELFARE CONCERNS

photo 4

2.1. Barren Living Quarters

Not only lay people agree, but also many biomedical and psychological researchers acknowledge that the barren cage is not an appropriate, let alone humane, permanent living environment for nonhuman primates who are:

1. social creatures, needing compatible companionship for their emotional well-being and behavioral health;

2. intelligent creatures, suffering from boredom and depression in a monotonous, relatively unchanging environment.

photo 5

Solitary imprisonment is a severe punishment for *human* primates.

photo 6

There is good reason to believe that solitary imprisonment is a terrible experience for *nunhuman* primates as well.

photo 7

The inadequacy of the barren cage is addressed by professional guidelines and legislative rules:

- The International Primatological Society (1993 & 2007) makes it very clear:
Pair or group housing in an enclosure must be considered the norm. For experimental animals, where housing in groups is not possible, keeping them in compatible pairs is a viable alternative social arrangement. Single caging should only be allowed where there is an approved protocol justification on ***veterinary or welfare grounds*** [emphasis added]. *Adequate space alone does not in itself provide for good welfare, but larger enclosures allow greater complexity of cage furnishngs and other enrichments, and greater flexibility for meeting social needs. The vertical dimension of the cage is of importance* [because of the vertical flight response] *and cages where the monkey is able to perch above human eye level are recommended. As animals like to work for their food, increasing processing time, increasing foraging, or providing puzzle feeders or other feeding devices is encouraged.*

- The Primate Research Institute (2003) of Japan underlines:
Primates are very social animals. Physical contact, such as grooming, and non-contact communication through visual, auditory, and olfactory signals are vital elements of their lives. Providing animals with a satisfactory social interaction helps to buffer against the effects of stress, reduce behavioral abnormalities, increase opportunities for exercise and helps to develop physical and social competence. Good relations between the animals and personnel is important for animals to reduce stress and for personnel to obtain safer working conditions. Personnel who have gained the trust of animals can more easily perceive abnormal behaviors and the animals are more likely to cooperate with them during research procedures, such as restraint and blood sampling. Devices suitable for gross motor and behavioral patterns, such as perches and three-dimensional structures should be arranged to make as much use of the available space as is possible. Diversity is essential to the housing environment of laboratory animals. Windows through which the animals can see the outside world may help to alleviate some boredom. Food presentation should satisfy the animal's interest in manipulating objects. In order to satisfy their requirement to interact with their environment, it is desirable to provide feeders that require complex handling or devices which in some way lead the animals to object manipulation.

- The United States Department of Agriculture (1991) stipulates:
Dealers, exhibitors, and research facilities must develop, document, and follow an appropriate plan for environmental enhancement adequate to promote the psychological well-being of nonhuman primates. The plan ***must*** [emphasis added] *include specific provisions to address the social needs of nonhuman primates of species known to exist in social groups in nature. The physical environment in the primary enclosures* ***must*** [emphasis added] *be enriched by providing means of expressing noninjurious species-typical activities. Examples of environmental enrichment include providing perches, swings, mirrors, and other increased cage complexities; providing objects to manipulate; varied food items; using foraging or task-oriented feeding methods; and providing interaction with the care giver or other familiar and knowledgeable person consistent with personnel safety precautions.*

- The National Research Council (1998) of the United States emphasizes:
Social interactions are considered to be one of the most important factors influencing the psychological well-being of most nonhuman primates. Knowing that most primates benefit from social interactions, it should be obvious that they can be harmed by a lack of social interaction. The common practice of housing rhesus monkeys singly calls for special attention. Every effort should be made to house these animals socially (in groups or pairs). Although the causes of self-directed biting are poorly understood, prolonged individual housing is probably an influential contributing factor. The animal technician's and caregiver's roles are pivotal to the social support of

primates, particularly animals that are singly caged. Under natural conditions, many primates spend much of their lives above ground and escape upward to avoid terrestrial threats. Therefore, these animals might perceive the presence of humans above them as particularly threatening. Even macaques, which some describe as semi terrestrial, spend most of the day in elevated locations and seek the refuge of trees at night. Optimal use of available cage space might well depend more on the placement of perches, platforms, moving and stationary supports, and refuges than on cage size itself.

- The Canadian Council on Animal Care (1984 & 1993) warns:
  Any primate housed alone will probably suffer from social deprivation, the stress from which may distort processes, both physiological and behavioural. In the interest of well-being, a social environment is desired for each animal which will allow basic social contacts and positive social relationships. Social behaviour assists animals to cope with circumstances of confinement.

- The National Health and Medical Research Council (1997) of Australia observes:
  For nonhuman primates *social interaction is paramount for well-being. Social deprivation in all its forms* ***must*** [emphasis added] *be avoided. Animals that need to be individually caged, either for experimental or holding purpose (for example, aggressive adult males), must be given contact with conspecific animals. Accommodation should provide an environment which is as varied as possible. It should meet the behavioural requirements of the species being used. Emphasis* ***must*** [emphasis added] *be placed on environmental enrichment.*

- The Medical Research Council (2004) of the United Kingdom requires:
  Primates ***must*** [emphasis added] *be provided with a complex and stimulating environment that promotes good health and psychological well-being and provides full opportunity for social interactions, exercise and to express a range of behaviours appropriate to the species. The volume and height of the cage (or enclosure) are particularly important for macaques and marmosets, which flee upwards when alarmed. Their cages and enclosures should be floor-to-ceiling high whenever possible, allowing the animals to move up to heights where they feel secure. Primates should be socially housed as compatible pairs or groups.* They *should not be singly housed unless there is* ***exceptional*** [emphasis added] *scientific or veterinary justification. Cages and enclosures should be furnished to encourage primates to express their full range of behaviours. Depending on the species, this should normally include provision for resting, running, climbing, leaping and foraging.* ***The MRC will require justification for the use of scientific procedures that restrict the opportunity to forage*** [emphasis added].

- The Council of Europe (2006) admonishes:
  Because the common laboratory non-human primates are social animals, they should be housed with one or more compatible conspecifics. Single housing should only occur if there is justification on ***veterinary or welfare grounds*** [emphasis added]. *The structural division of space in primate enclosures is of paramount importance. It is essential that the animals should be able to utilise as much of the volume as possible because, being arboreal, they occupy a three-dimensional space. To make this possible, perches and climbing structures should be provided.*

- The National Center for the Replacement Refinement Reduction of Animals in Research (2006) recommends:
  Primates should be socially-housed as compatible pairs or groups. Cages and enclosures should be floor to ceiling high whenever possible, with adequate perching to allow all animals to move up to heights where they feel more secure. The vertical and horizontal dimensions of the cage and enclosure should be exploited fully by incorporating shelves, logs, ladders, climbing structures, branches, hammocks, swings, ropes and objects to manipulate. All primates should be given the opportunity to forage daily, by scattering food in litter or substrate on the floor, or in a tray, and by using devices that encourage foraging activity.

2.2. Involuntary Restraint

Nonhuman primates—just like human primates—are sensitive creatures who do not want to be restrained against their will, but experience intense fear when they are forcibly subjected to life-threatening treatment.

It may be true that procedures such as injection and blood sampling are *simple,* but they can be *expected to produce little or no discomfort* (Scientists Center for Animal Welfare, 1987) **only** if the subject is **not** forced to leave her or his cage and subsequently is **not** forced to hold still during such a procedure. A needle prick is not a big deal, but the coercive contact with the human predator is a most distressing experience for any nonhuman primate.

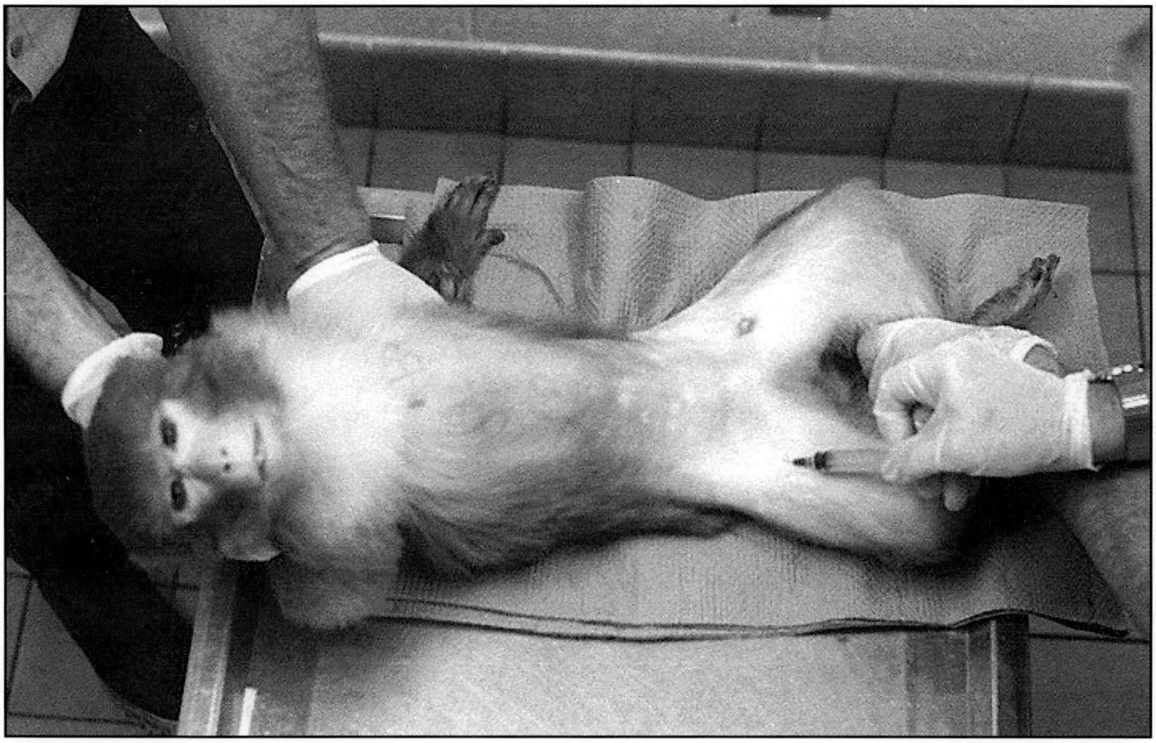

photo 8

The inadequacy of involuntary restraint is addressed by professional guidelines and legislative rules:

- The International Primatological Society (2007) reminds:
  Primates of many species can be quickly trained using positive reinforcement techniques to cooperate with a wide range of scientific, veterinary and husbandry procedures. Such training is advocated whenever possible as a less stressful alternative to traditional methods using physical restraint. Techniques that reduce or eliminate adverse effects not only benefit animal welfare but can also enhance the quality of scientific research, since suffering in animals can result in physiological changes which are, at least, likely to increase variability in experimental data and, at worst, may even invalidate the research. Restraint procedures should be used only when less stressful alternatives are not feasible.

- The Primate Research Institute (2003) of Japan warns:
  Physical or chair restraint, most definitely affects the behavior and psychology of laboratory animals. All possible measures to reduce their incidence should be taken. Animals should be trained to be as cooperative

as possible to the procedures to facilitate the rapid completion of work and to alleviate stress in both the animals and people in charge.

- The National Research Council (1998) of the United States observes:
  Procedures that reduce reliance on forced restraint are less stressful for animals and staff, safer for both, and generally more efficient. To reduce the stress of physical restraint, many primates can be trained for routine procedures.

- The Public Health Service (1996) of the United States recommends:
  Unless the contrary is established, investigators should consider that procedures [such as enforced restraint during life-threatening procedure] *that cause pain or distress in human beings may cause pain or distress in other animals.*

- The Home Office (1989) of the United Kingdom points out:
  The least distressing method of handling is to train the animal to co-operate in routine procedures. Advantage should be taken of the animal's ability to learn.

- The Medical Research Council (2004) of the United Kingdom states:
  Positive reinforcement techniques should be used to train primates to cooperate with catching, handling, restraint and research procedures. The routine use of squeeze-back cages and nets should be actively discouraged.

- The Canadian Council on Animal Care (1993) stipulates:
  Restraint procedures should only be invoked after all other less stressful procedures have been rejected as alternatives. Physiological, biochemical and hormonal changes occur in any restraint animal and investigators should consider how these effects will influence their proposed experiments.

- The Council of Europe (2006) underlines:
  Primates dislike being handled and are stressed by it; training animals to co-operate should be encouraged, as this will reduce the stress otherwise caused by handling. Training the animals is a most important aspect of husbandry, particularly in long-term studies.

Enforced restraint is sometimes advocated with the assertion that nonhuman primates are unpredictable and readily scratch and bite handling personnel (Gisler et al., 1960; Ackerley and Stones, 1969; Valerio et al., 1969; Altman, 1970; Whitney et al., 1973; Henrickson, 1976; Wickings and Nieschlag, 1980; Robbins et al., 1986; Wolfensohn and Lloyd, 1994; Johns Hopkins University and Health System, 2001; Panneton et al., 2001; University of Arizona - IACUC Certification Coordinator, 2008; University of Minnesota - Investigators, and Animal Husbandry and Veterinary Staff, 2008). This precautious warning overlooks the fact that the animals are not intrinsically "aggressive," but that enforced restraint *makes* them aggressive. Trying to bite or scratch the handling personnel is the biologically normal self-defense of any animal who is forcibly restrained by a natural predator. The very act of forceful restraint triggers, rather than prevents, aggressive self-defense. Gaining the animal's trust, and then training him or her to cooperate, during procedures eliminates the risks that are associated with self-defensive aggression. A cooperative animal is no longer given any reason to bite or scratch the investigator, animal technician or veterinarian who is working *with*, rather than against, the animal during a procedure.

3. REFINEMENT

3.1. Animate Enrichment

Animate enrichment promotes non-injurious contact and interaction with one or several compatible conspecifics or with humans whom the animal can trust.

photo 9

Given that nonhuman primates are social—not solitary—animals, it is logical that their need for compatible companionship *must* be addressed when they are imprisoned in research laboratories.

Partner-directed grooming is probably the most important social behavior of nonhuman primates. Studies of wild populations have shown that Old World primates spend 5 to 25 percent of the day interacting with each other, with grooming being the prevalent social activity (Hall and De Vore, 1995; Lindburg, 1971; Teas et al., 1980; Chopra et al., 1992; Wrangham, 1992; Leon et al., 1993; Hanya, 2004; McNulty et al., 2004); corresponding data of New World primates are missing.

photo 10

When grooming each other, the animals give the impression of being absorbed in the interaction. The recipient of grooming, in particular, leaves no doubt that she or he finds being groomed pleasurable and relaxing. There is scientific evidence that grooming serves to reduce tension and stress in the passive partner, and perhaps also in the active partner of this interaction (Terry, 1970; Schino et al., 1988; Boccia, 1989a; Boccia et al., 1989; Keverne et al., 1989; Gust et al., 1993; de Waal and Aureli, 1997; Aureli et al., 1999; Judge et al., 2006; Shutt et al., 2007). The animal in the barren single-cage is deprived of this very positive experience.

3.1.1. Group-Housing

Housing nonhuman primates in compatible groups of three or more animals of both sexes would be the optimal strategy to address their social needs.

3.1.1.1. Group Formation

There are numerous reports on integrating animals into already established core groups, but only a few reports on forming a new group of previously single-caged individuals.

Fritz and Fritz (1979) and Fritz (1994) developed a protocol to introduce previously single-caged chimpanzees (*Pan troglodytes*) to unfamiliar peers. The newcomer is first moved into a specially designed social unit and kept next to the cage of a selected member of an already established group. The two chimpanzees have full olfactory, visual and auditory contact, as well as limited tactile contact. The selected group member is moved in as a cage mate for the newcomer as soon as friendly interactions through the separating cage mesh are consistently observed. After several days, another group member is introduced to the pair in this same way, then another is introduced to the trio, and so on until the newcomer has met all members of the group and is then fully integrated. A total of 59 of 60 chimpanzees of both sexes and all age classes were successfully socialized to compatible group-living in this manner, without a single incidence of serious fighting.

photo 11

photo 12

Kessel and Brent (2001) tranquillized adult single-caged baboons (*Papio* spp.) with ketamine and placed one trio of males in one enclosure and two trios of two females and one male in two other enclosures, where the animals regained consciousness in their respective groups. The formation of the three groups was accompanied by two incidences of wounding, which were superficial and required no medical treatment. All three trios were compatible and remained stable. Bourgeois and Brent (2005) confirmed these findings in a subsequent study with three adolescent male baboons. Group formation was accompanied by no overt aggression. Rough-and-tumble wrestling was observed and dominance was quickly established with all agonistic encounters followed by grooming.

Bernstein and Mason (1963) released 11 rhesus macaques (*Macaca mulatta*)—three adult females, two adult males, one subadult female, one subadult male and four juveniles—simultaneously into a large enclosure. During the first hour, a total of 83 *threats* and 23 *attacks* were observed; injurious encounters were not recorded, but one of the two males *soon showed signs of deteriorating health and died after 20 days.*

Reinhardt (1991a) tried to form an isosexual group of six previously single-caged adult female and another group of six previously single-caged adult male rhesus macaques. Future group members were first given ample opportunity to physically interact with each other on a one-to-one basis during a one-week period. Dominance-subordinance establishment was ascertained in each dyad. The two groups were then formed by releasing the six animals simultaneously into a big cage. In both situations, aggressive incompatibility was heralded by certain subjects challenging other partners to whom they had been subordinate during the familiarization week. Aggressive harassment was intense and persistent. Alliances were quickly formed and several animals in union attacked selected targets. Victims were cornered, and they showed no resistance, except for fear-grinning and submissive crouching; they did so to no avail and the vicious attacks continued. Both groups were disbanded within the first hour to avoid fatal consequences.

Gust et al. (1991) introduced eight unfamiliar adult female rhesus macaques together with one unfamiliar adult male simultaneously in a large enclosure. There was no serious fighting, and in fact no contact aggression was recorded, even though firm dominance-subordinance relationships were established during the first 48 hours. Several females stayed in close proximity of the male, who copulated with two of them during the first day. The male's presence accounted for the females' tolerance of each other.

Gust et al. (1996) duplicated this study with eight adult female and one adult male pig-tailed macaques (*Macaca nemestrina*) with the same results: Group formation and the establishment of a social hierarchy was not associated with serious aggression; there was no contact aggression during the first five hours following the simultaneous release of the eight animals into the same enclosure.

Clarke et al. (1995) familiarized three single-caged adult male long-tailed (cynomolgus) macaques (*Macaca fascicularis*) pairwise with each other in a non-contact housing arrangement for two weeks and subsequently released them as a trio in a large cage. No injurious fighting was recorded; the new group was compatible.

Asvestas (1998) and Asvestas and Reininger (1999) established a group of 22 adult male long-tailed macaques by first forming 11 compatible pairs. After nine months, all animals were sedated with ketamine and placed simultaneously in a big enclosure where they regained consciousness under careful supervision of the attending staff. The new group turned out to be compatible, even though four males were slightly injured during fighting.

Clarke et al. (1995) kept three lion-tailed macaques (*Macaca silenus*) in a housing arrangement that allowed all animals to see each other for a period of two weeks. The three males were subsequently released simultaneously into a large cage. This event was not accompanied by serious fighting, but the group was disbanded because the three males avoided each other and were apparently sufficiently distressed that their well-being was compromised, especially that of the lowest ranking animal, who did not obtain sufficient food.

Stahl et al. (2001) released six unfamiliar adult lion-tailed macaques into well-structured large living quarters and encountered no aggression-relation problems. The six males showed no contact aggression, but 91 non-contact agonistic interactions during the first six hours.

King and Norwood (1989) released 11 single-caged female and 13 single-caged male squirrel monkeys (*Saimiri* spp.), ranging in age from 1 to 18 years, without any preliminaries, into a well-structured room. The establishment of the new group was accompanied by two deaths—one male and one female—resulting from attacks by other monkeys.

No foolproof recipe is yet available for group formation [of capuchin monkeys]. *Our knowledge of how to form or modify capuchin groups (Cebus spp.) does not come from systematic experimental study, but derives from husbandry problems faced occasionally by laboratories. Overall, group formation is a stressful procedure both for the animals and the care-givers, and although cumulative experience may help to reduce the risks of failure, the outcome can never be predicted with absolute certainty* (Visalberghi and Anderson, 1999).

3.1.1.2. Group-Housing

3.1.1.2.1. Behavioral Health

Alexander and Fontenot (2003) established isosexual male groups (average group size four animals) of 80, previously single-caged adult male rhesus macaques. Thirty-one (39 percent) of these males had at least one prior incidence of self-injurious biting (SIB). During the year prior to group formation, the clinical history of the subjects included a 20 percent incidence of diarrhea and a 13 percent incidence of SIB requiring veterinary care. During the first four months after group formations less than 2 percent of the animals suffered from diarrhea, and no animal showed signs of SIB.

Fritz (1989) transferred four individually housed chimpanzees who engaged in self-mutilation to compatible group-housing arrangements. The behavioral pathology gradually ceased in all four subjects.

3.1.1.2.2. Problems

Group-housing in the research laboratory setting can bear substantial risks for individual members of the group, especially when mature animals of both sexes are present. The inherent constraints of confinement often make it impossible for individuals to keep appropriate social distance from each other, so as to avoid conflicts. Research-related and management-related interferences in the group's membership are bound to destabilize its social structure, thereby triggering rearrangements in the social hierarchy that are usually associated with overt aggression.

photo 13

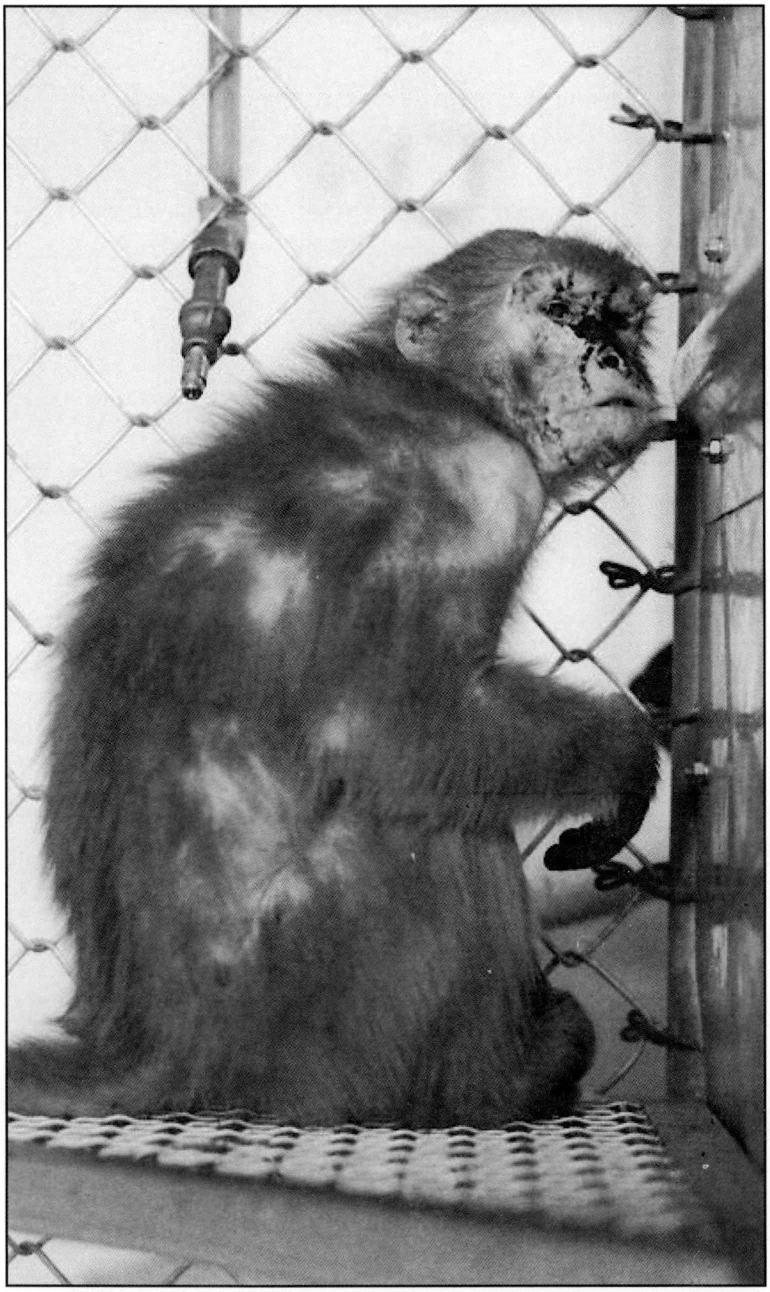

photo 14

Serious, sometimes fatal injuries resulting from aggression are not uncommon in captive groups of baboons (Rowell, 1967; Nagel and Kummer, 1974), pig-tailed macaques (Sackett et al., 1975; Erwin, 1977), rhesus macaques (Kaplan et al., 1980; Kessler et al., 1985; Schapiro et al., 1994), squirrel monkeys (Abee, 1985), marmosets (Poole, 1990), chimpanzees (Alford et al., 1995), and vervet monkeys (Knezevich and Fairbanks, 2004).

No published report could be found of serious aggression problems in core groups of long-tailed macaques (cf., Aureli et al., 1993; Clarke et al., 1995; Ljungberg et al., 1997), stump-tailed macaques, mangabeys, capuchin monkeys (cf., Fragaszy et al., 1994), and tamarins (cf., Poole et al., 1999).

3.1.2. Pair-Housing

To enhance the life-style of a primate, one of the most effective, but often overlooked improvements is pair housing (Rosenberg and Kesel, 1994). Keeping nonhuman primates in compatible pairs is a good compromise to group-housing; it addresses the animals' basic social needs while providing more assurance of their safety, better access to individuals, and control over their reproduction.

Initial, strong reservations against the transfer of single-caged animals to pair-housing arrangements have proven to be based on the erroneous idea of the *aggressive and near-intractable monkey* and the disregard of basic ethological principles when establishing new pairs.

photo 15

photo 16

3.1.2.1 Pair Formation

3.1.2.1.1. Introducing Juveniles to Adults

Adults—both females and males—are normally inhibited from showing overt aggression toward juveniles. This circumstance makes it easy to transition single-caged adults to compatible pair-housing arrangements: the naturally weaned juvenile is simply introduced to the adult in the adult's home cage. Typically, the adult will show parental responses, huddling with the young, spending much time grooming the young, and allowing the young to engage in often exuberant play behaviors. Even rhesus males, who have the reputation of being particularly aggressive, have the tendency to treat their little companions with gentleness and great tolerance.

photo 17

photo 18

photo 19

Reinhardt (1994a) transferred naturally weaned, 12 to18 months old surplus infants from a rhesus macaque breeding colony without any preliminary precautions, pairwise to unfamiliar single-caged adults of both sexes. A total of 78 pairs were tested and pair compatibility ascertained during the first week in 96 percent (75/78) of cases:

- the adult did not injure the juvenile,
- the juvenile showed no signs of depression, and
- the adult shared food with the juvenile.

Three pairs (4 percent) were incompatible. One female grabbed her female infant immediately upon her arrival; she continued to do this repeatedly during the next 30 minutes, after which the infant was removed. One male bit his infant on the fourth day of introduction. The youngster was slightly injured, although not bleeding. When the infant started to consistently avoid the adult, the pair was split. Another male often grabbed his infant companion, even though he gently groomed him, and the two huddled with each other regularly. Gradually, however, the infant showed more and more avoidance behavior, and the two were finally separated after nine days.

photo 20

3.1.2.1.2. Introducing Juveniles to Juveniles

Juveniles who have not yet reached the age when they become ambitious to dominate over others are usually compatible when they are introduced as pairs, even when they are strangers to each other.

Reinhardt (1994a) transferred a total of 84 female and 22 male juvenile rhesus macaques to same-sex pair arrangements. All 42 female and all 11 male pairs were compatible throughout a one-year follow-up period. Males were occasionally observed playfully wrestling with each other, but this never resulted in injurious aggression or depression.

photo 21

photo 22

photo 23

photo 24

3.1.2.1.3. Introducing Adults to Adults

Adult primates have the tendency to react with hostility when they meet another adult conspecific with whom they are not familiar. Strangers first determine their dominance-subordinance relationship which often involves fighting. To avoid this in the laboratory setting, adults assigned to be paired are first given the opportunity to get to know each other by being kept in a double cage (photo 23) where they can settle their relationship via non-contact communication through a grated or transparent cage divider (photos 24 and 25).

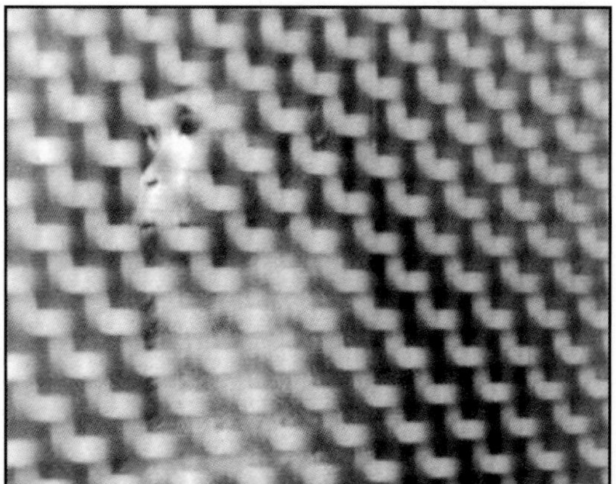

photo 25

Doyle et al. (2008) biotelemetry-instrumented eight single-caged adult male rhesus macaques, carefully familiarized the animals with each other, and then introduced them pairwise. All four pairs were compatible. Upon introduction, subjects showed no increased heart rate, indicating that the pair formation process was not a stressful experience for them.

Compatible companions typically engage in affiliative, rather than aggressive, interactions when they are introduced to one another as a new pair.

photo 26

Reinhardt et al. (1988) and Eaton et al. (1994) familiarized previously single-caged adult rhesus females in double cages with transparent partitions for one week, and then introduced them as pairs in a different double cage. Within the first two hours after introduction, dominance-determining fighting was witnessed in 27 percent (5/18) and 10 percent (2/21) of cases, respectively. The fights resulted in no serious injuries, but they were persistent and resulted in depression in the victim in three and two dyads, respectively. These five pairs were classified as incompatible and the partners were permanently separated. Consequently, pair compatibility during the first week was 83 percent (15/18) and 90 percent (19/21), respectively.

Abney and Weed (2006) familiarized 56 adult male rhesus macaques with an unspecified number of adult, subadult and juvenile males in a grooming-contact bar housing arrangement, but did not check whether potential partners established dominance-subordinance relationships. When the dividing bars were subsequently removed, only 61 percent (34/56) of the pairs were compatible. Serious fighting or injuries occurred in 39 percent (22/56) of the dyads; this prompted the authors to warn that *potentially dangerous socializations can occur under laboratory conditions.*

Watson (2002) removed the mesh panels between familiarized adult long-tailed macaque males when the two neighbors showed no *aggressive activity*. Of 31 pairs formed in this way, two (6 percent) had to be separated and treated within the first four hours because *one animal in each pair sustained injuries during minor fighting.*

Reinhardt (1994a) made sure that the partners of 77 adult female rhesus dyads and 20 adult male rhesus dyads had established their dominance-subordination relationships during a non-contact familiarization period, before they were introduced in a different double cage. This precaution was implemented in order to minimize the animals' need to engage in dominance-determining aggression upon being introduced with each other. The occurrence of strictly unidirectional

- fear-grinning when being looked at by the neighbor,
- withdrawing and/or looking away when being approached or looked at by the neighbor, and
- enlisting against other animals of the room or against the observer

were taken as indicators that one animal was subordinate and accepted the dominant position of his or her neighboring partner. Partners who had established such a relationship were then introduced to each other in a different (to avoid potential territorial antagonism) double cage. Newly formed pairs were regularly observed during the first week.

Shortly after introduction, fighting took place in only two of the 97 days tested.

photo 27

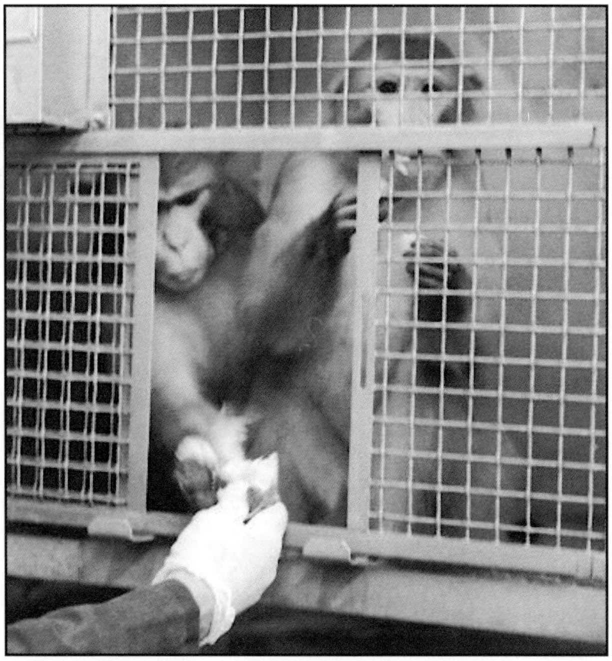

photo 28

Partners turned out to be compatible in 95 percent (73/77) of the female pairs and also in 95 percent (19/20) of the male pairs. Compatible partners did not engage in serious aggression, they shared food—both standard food and supplemental food—and none of them became depressed.

Roberts and Platt (2005), Byrum and St. Claire (1998) and Lynch (1998) applied the same pair formation technique with 13 adult male rhesus dyads, 12 adult female pig-tailed macaque dyads, and 17 adult male long-tailed macaque dyads, respectively. Potential pairs had all established clear-cut dominance-subordinance relationships prior to partner introduction. Partner introduction was accompanied by fighting in only 2 percent of the 42 dyads tested. Pair compatibility was:

- 92 percent (12/13) for the male rhesus macaques,
- 100 percent (12/12) for the female pig-tailed macaques, and
- 94 percent (16/17) for the male long-tailed macaques.

Crockett et al. (1994) also non-contact familiarized the potential partners of 15 adult male and 15 adult female long-tailed macaques, but introduced the animals as pairs in the familiarization cage without prior verification that they had established dominance-subordination relationships. Under these circumstances, fighting occurred shortly after partner introduction in 67 percent (10/15) of the male pairs and in 13 percent (2/15) of the female pairs. Over the course of the first week, 80 percent (12/15) of the male pairs and 100 percent of the female pairs turned out to be compatible.

Reinhardt (1994) transferred 10 adult female and six adult male stump-tailed macaques from single-housing to isosexual pair-housing by first allowing potential partners to establish dominance-subordinance relationships without risk of injury during a non-contact familiarization phase. Following subsequent introduction in a new home cage, all eight pairs showed signs of compatibility. Female partners reconfirmed their rank relationships within 30 minutes with subtle gestures, never by overt aggression. Male partners engaged in hold-bottom rituals (de Waal and Ren, 1988) upon being introduced to each other. Two male pairs reconfirmed rank relationships within 30 minutes with gestures, while the third pair resorted to a brief non-injurious dominance-reconfirming fight, which was followed by another reconciliatory hold-bottom ritual.

photo 29

Coe and Rosenblum (1984) introduced 10 adult unfamiliar male bonnet macaques (*Macaca radiata*) pairwise without any preliminaries. *As usually occurs when unfamiliar males first meet, agonistic behaviors related to the establishment of dominance relations occurred at pair formation. The aggressive incidents were limited, usually involving threats and pursuit behavior, and manual attacks occurred only infrequently. More typically, one animal submitted and indicated his subordinate status through communicative gestures. In the first week following pair formation, the occurrence of aggressive behavior subsided almost entirely.* The males' response to this pairing procedure may reflect their reputation of possibly showing *the highest degree of male-male tolerance in the genus Macaca.*

Bourgeois and Brent (2005) established four pairs of previously single-caged subadult male baboons by sedating potential companions and having them wake up together in the same cage. No serious aggression was witnessed during 10 half-hour observations conducted during the first two weeks.

Jerome and Szostak (1987) allowed an unspecified number of adult female baboons to live pairwise with each other four hours a day, three times a week. *The same pairs visited each other in either animal's cage. No significant aggression occurred during visits.*

We have never managed to house adult male vervet monkeys (*Cercopithecus* spp.) *in pairs, unless they were reared together right after weaning, in which case pair compatibility is about 90 percent* (LAREF, 2007a).

Majolo et al. (2003) checked the clinical records of 56 unfamiliar female marmosets (*Callithrix* spp.) of different age classes who were paired with each other without prefamiliarization. The animals engaged in considerable aggression and only 79 percent of the 28 pairs were allowed to stay together beyond the first week; six pairs (21 percent) *were split up because one of the monkeys was subject to intense aggression and/or was seriously injured as a consequence of fighting.*

photo 30

3.1.2.2. Pair-Housing

3.1.2.2.1. Long-Term Pair Compatibility, Behavioral Health, Practicability, Physical Health

Reinhardt (1994a) formed 84 compatible pairs of juvenile female and 22 compatible pairs of juvenile male rhesus macaques and noted that the animals remained compatible for at least 12 months. There were 21 juvenile female pairs with cranial implants. Living together in the same cage did not constitute any specific risks for the animals (no local infections possibly caused by grooming the margins of the implantation site) and no risk for the implants (no damage related to social interactions).

photo 31

photo 32

Reinhardt (1994a) established 75 compatible adult-infant pairs who were allowed to stay together uninterruptedly. Compatibility was ascertained throughout a 12-month follow-up period. Incompatibility was noted after more than one year in two cases, when the now prepubertal young subjects started teasing their over 30 years old companions, thereby creating excessive disturbance for these aged animals. Two of the infants lived with adult females who were tethered, and 32 paired infants had cranial implants. Both circumstances did not interfere with research protocols requiring remote sample collection and neuroendocrinological testing.

photo 33

Crockett et al. (1994) established 15 compatible adult female and 12 compatible adult male long-tailed macaque pairs and housed them in such a way that partners were separated each day for 17 hours and subsequently reunited for 7 hours. While 100 percent of the female pairs successfully coped with this situation and remained compatible, only 50 percent of the male pairs adjusted. The other 50 percent became incompatible and had to be separated within two weeks of living together under these socially challenging conditions.

Lynch (1998) also formed 16 compatible adult male long-tailed macaque pairs, but partners could stay together without interruption. All pairs remained compatible throughout a 12-month follow-up period and longer.

Coe and Rosenblum (1984) observed five adult male bonnet macaque pairs on four different days during 15-minute sessions in the course of the first week after the pairs were established. Subjects groomed and contacted each other on average 29 percent of the time.

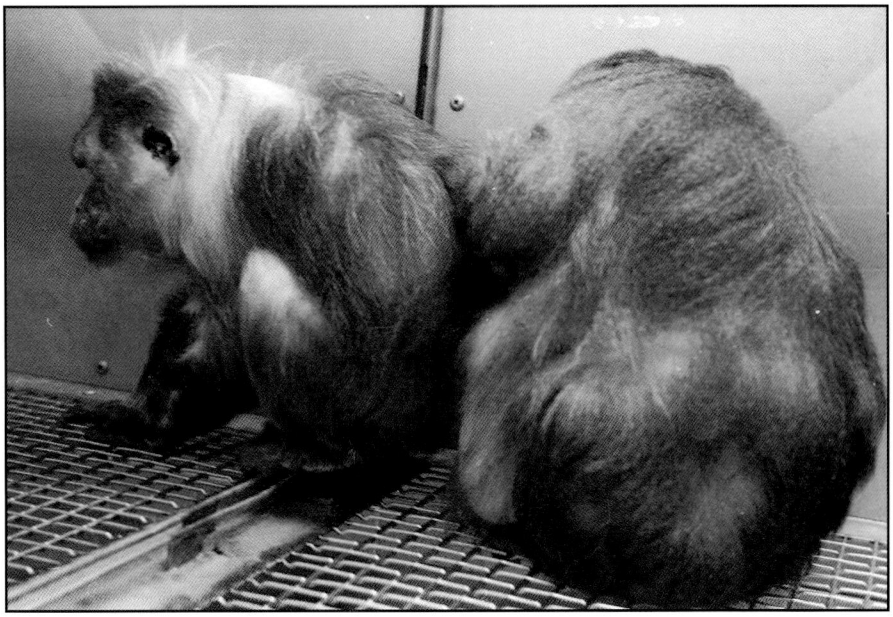

photo 34

Reinhardt (1994b) monitored five adult female and three adult male stump-tailed macaque pairs, who had lived together for six months, each pair for one hour. On average, females groomed each other 19 percent of the time and hugged each other 6 percent of the time; males groomed each other 13 percent of the time and engaged in holding bottom rituals 4 percent of the time.

photo 35

Line et al. (1990a) observed 10 adult female long-tailed macaques, who lived in five compatible pairs, daily over a period of two weeks for a total of 18 hours. The animals groomed each other, on average, 31 percent of the time.

Crockett et al. (1994) recorded the behavior of 15 female and 8 male pairs of adult long-tailed macaques during 90-minute test sessions, 13 days after the pairs were formed. On average, female companions groomed each other 35 percent of the time; male companions groomed each other 17 percent of the time (difference is statistically significant).

Doyle et al. (2008) allowed eight single-caged adult male rhesus macaques to live uninterruptedly as four compatible pairs. Over the course of 18 months, one bite laceration was incurred [after 3.5 months], but the pair remained compatible after the injury was treated and healed. Average fecal cortisol levels were significantly higher when the males lived alone than after they had lived with a companion for 20 to 39 weeks (83 ng/g versus 9 ng/g), indicating that long-term pair-housing was a less distressing situation than single-housing.

photo 36

Behavioral scan sampling revealed that the males groomed each other about 13 percent of the time.

photo 37

photo 38

Eaton et al. (1994) studied 12 newly formed, compatible adult female rhesus pairs over a 36-month period. During this time, only one pair became incompatible, when the two partners had a serious fight. Systematic 10-minute observations were carried out on the 22 long-term compatible animals at the time when they were still caged alone and, again, when they had lived with their cage mates for six months. These animals spent about 2 percent of the time pulling their hair when they were single-housed, versus only 0.3 percent of the time when they were pair-housed; the difference was statistically not significant. When they lived with a companion they groomed each other about 31 percent of the time.

photo 39

photo 40

Reinhardt (1990a & 1994a) formed 73 compatible adult female and 19 compatible adult male rhesus macaque pairs. The animals were allowed to live together uninterruptedly. Over a 12-month follow-up period, compatibility was 93 percent (68/73) for the female pairs and 84 percent (16/19) for the male pairs. During two 30-minute video recordings of eight female pairs and four male pairs:

Females, on average, groomed each other 25 percent and hugged each other 4 percent of the time; Males, on average, groomed each other 12 percent and hugged each other 2 percent of the time; the sex difference was statistically significant.

Among the compatible pairs, were four 30 to 35 years old animals who were so old that they experienced a progressive loss in body weight. Living with a companion did not accelerate this biological process, indicating that the permanent presence of a companion did not jeopardize their general health. These senile rhesus macaques groomed each other, on average, 21 percent of three one-hour observations (Reinhardt and Hurwitz, 1993).

photo 41

Some animals were assigned to controlled food intake studies over the course of the first two years after pair formation. When this happened, they were allowed to stay in their home cage, where they were separated from their companions with a grated cage dividing panel during the day, and reunited for the night after food intake was recorded.

photo 42

The majority of the animals were assigned to a timed breeding program. All 18 females who gave birth during the first two years after pair formation were allowed to stay with their partners. The presence of offspring did not affect the compatibility between the two cage companions.

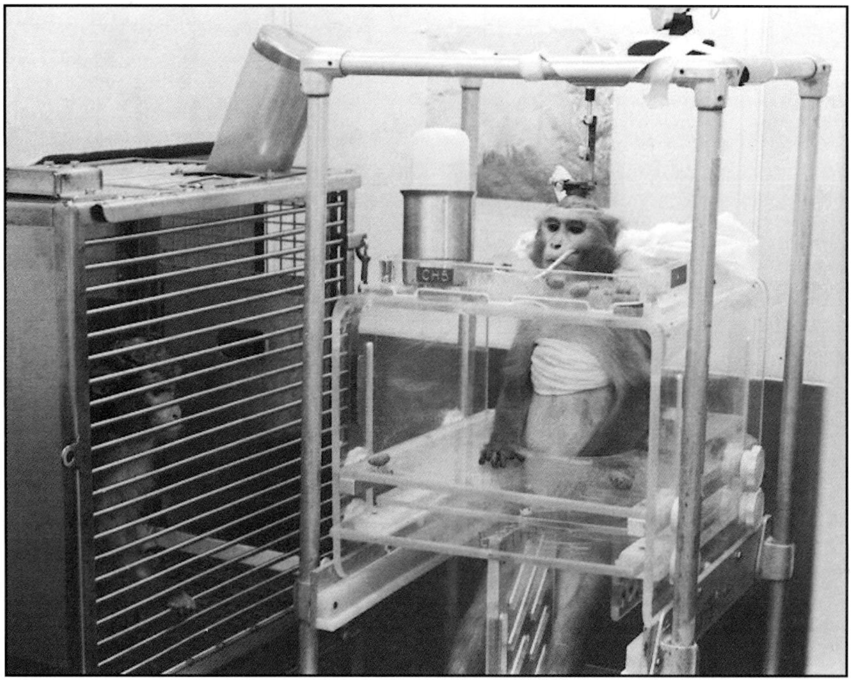

photo 43

There were 23 female pairs with one or both partners having cranial implants. This circumstance did not jeopardize the integrity of ongoing neurophysiological research of one or both animals. When one partner had to be chair-restrained during an experiment, the companion was brought along in a mobile cage to provide emotional support.

photo 44

Roberts and Platt (2005) studied one adult male long-tailed and eight adult male rhesus macaques who all had cranial implants and lived with compatible partners in a pair-housing arrangement. The presence of a social partner did not cause any problems with the implants, which lasted for an average of 21 months. Partners were separated daily for a few hours to participate in physiological experiments; this had no adverse effect on their compatibility which, depending on the length of the study, could be confirmed for up to 40 months.

Murray et al. (2002) demonstrated the practicability of post-operative pair-housing in 15 female long-tailed macaques who were returned to their partners on the day of the operation. Change in dominance status, self-traumatic events, weight loss or diarrhea did not occur in any of these animals, and the incision sites healed unremarkably. The animals ate and drank normally, and they accepted their post-operative oral medication.

Our long-tailed macaques are subjected to a lot of orthopedic procedures. There have never been problems with the re-pairing of the animals after surgery. We partition the pair's cage with a transparent panel, which we remove after the treated companion has fully recovered from anesthetic effects (usually 24 hours). It has never happened that animals who had no surgery showed any negative behavioral reactions toward their temporarily probably weaker cage mates (LAREF, 2007b).

Line et al. (1990) transferred five adult female long-tailed macaques who engaged in self-injurious behavior to compatible pair-housing arrangements. *Self-abusive behaviors were completely absent after pair formation.*

Reinhardt (1999) worked with three adult female and four adult male rhesus macaques who habitually bit themselves when they were caged alone. The provision of perches, gnawing sticks and food puzzles did not alleviate this behavioral pathology, but when the seven animals were successfully paired with compatible partners, the self-biting stopped immediately in three cases and gradually in the remaining four cases.

Weed et al. (2003) vasectomized six single-caged rhesus males who engaged in persistent self-injurious biting and paired them with adult females. Three of these males stopped the self-biting after being transferred to social-housing, and self-biting was no longer noticed during a one to six months follow-up period. Socialization had a moderating, but not healing, effect in the other three males.

Reinhardt (1990b) assessed the clinical records of a rhesus macaque colony consisting of 237 single-housed and 382 pair-housed animals of both sexes and all age classes. The incidence of non-research-related veterinary treatment was 23 percent for single-caged animals, versus 10 percent for pair-housed animals, indicating that the animals' physical health was not jeopardized by sharing a cage with a companion.

Schapiro and Bushong (1994) examined the clinical records of 98 juvenile rhesus macaques during one year when they were caged alone and the subsequent year when they lived in opposite-sex pairs. Individuals required veterinary treatment more than twice as often when they were single-housed (0.40 times/year) than when they were pair-housed (0.17 times/year).

3.1.2.2.2. Social Buffer Effect

It has been demonstrated in pair-housed squirrel monkeys and rhesus macaques that the two companions do not differ in serum cortisol concentration and immune response (Coe et al., 1982; Gonzalez et al., 1982; Reinhardt et al., 1991; Eaton et al. 1994). Rather than being a source of distress, the compatible companion can serve as a social buffer during potentially stressful research-related situations, such as being chair-restrained in a test room.

Gust et al. (1994) transferred seven adult female rhesus monkeys from their group to an unfamiliar environment, either alone or together with a group member. During both conditions, subjects were initially equally distressed, as measured in alterations of cell-mediated immune parameters, but they recovered significantly quicker when they had the social support of a companion.

Mason (1960) and Gunnar et al. (1980) placed five and 12 infant rhesus macaques into a strange environment, either alone or as a pair with another infant. Subjects showed significantly fewer signs of distress (crouching, self-clasping, vocalization, agitation) when they were tested in the company of another monkey, indicating that the companion had a calming, reassuring influence.

Similar observations were made by Hennessy (1984) in eight squirrel monkey infants, who vocalized significantly less when they were tested in an unfamiliar environment as a pair than when they were tested alone. A significant elevation of plasma cortisol was observed only when the animals were exposed to the novel environment alone, not when a companion was present. Coe et al. (1982) noticed the same stress reducing effect in 14 adult male squirrel monkeys. Subjects showed significantly fewer distress reactions (vocalization, fear reactions, agitation) to a snake behind a mesh wall when another male was with them than when they were confronted with the snake alone.

Gonzalez et al. (1982) exposed six single-housed and six pair-housed adult female squirrel monkeys to the stress of capture followed by anesthesia and cardiac puncture. The 30-minute plasma cortisol increase was significantly lower in subjects housed with a companion than in subjects housed alone (38 versus 60 percent).

Coelho et al. (1991) measured blood pressure via arterial catheter implants of four tethered adult male baboons who were kept in a test room alone or in company of a familiar male baboon with whom they had visual, tactile and auditory contact through a wire mesh panel. Mean blood pressures were significantly lower when another baboon was present suggesting that companionship mitigated the distress response to the unfamiliar test room environment.

photo 45

3.1.3. Grooming-Contact Housing

Crockett et al. (1997) housed same-sex pairs of adult long-tailed macaques in double-cage units in which partners were separated by a blind panel for 19 hours daily. During the remaining five hours of the 24-hour day, they were separated by *grooming-contact bars*, allowing them to reach through with their arms. Of 16 female pairs tested, 100 percent were compatible and partners spent about 43 percent of the time grooming each other. Of 45 male pairs tested, 89 percent were compatible and partners spent about 7 percent of the time grooming each other.

The usefulness of grooming-contact bars, or woven wire panels with mesh openings large enough so that adjacent neighbors can groom each other (Coelho and Carey, 1990), has also been confirmed in adult iso- and heterosexual pairs of baboons (Coelho et al., 1991; Crockett and Heffernan, 1998) and adult heterosexual pairs of pig-tailed macaques (Crockett et al., 2001; Lee et al., 2005).

Compared with other species, rhesus macaques do not adjust well to the grooming-contact housing system (Crockett et al., 2006).

3.1.4. Kindergarten

Weaned macaque infants are often raised alone or in pairs. *It is my experience with rhesus macaques that an optimal environment for these youngsters is a kindergarten in which one adult animal "keeps order" and shows quasi-parental affection* (Reinhardt, 2008).

photo 46

photo 47

I always enjoyed observing the animals. The youngsters spent very much time of the day playing with each other socio-sexually and grooming the nanny. The young males often engaged in wrestling games that typically served as a prelude to extended grooming sessions (Reinhardt, 2008).

photo 48

photo 49

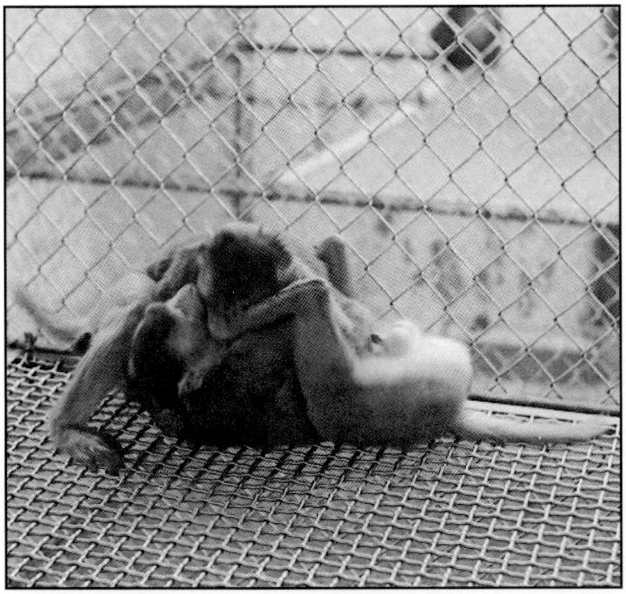

photo 50

photo 51

photo 52

Older juveniles, both females and males, would typically carry around little infants and protectively cradle them. They developed strong affectionate bonds with each other which we respected when the time came to remove prepubertal animals and transfer them to same-sex pair-housing arrangements (Reinhardt, 2008).

I used to do something similar when we weaned our infants. We transferred them into a pen with a big brother or big sister whose function was to "teach the youngsters how to become adults." The trick was finding a good "aunt" or "uncle" for the kids. We re-used those adults every year when we weaned infants. It was a great way to socially house some of our retired rhesus males and females. I think they accepted their assigned role, because the "uncle" or "aunt" would have a whole army of kids to groom her or him (imagine one big male rhesus being groomed by 5-6 weanlings!). These adults were very tolerant of the kids, but they also taught them boundaries. In the beginning, they would allow the weanlings to cling to them, "steal" their food and do all kind of antics, but as time went on you could see them setting more and more limits; but they still remained very tolerant. When the youngsters started to become sexually active, we formed same-sex groups or new breeding troops (Murphy, 2008).

photo 53

3.1.5. Attention from Care Personnel

Captive primates, especially animals living in small cages, often show intense fear reactions when a person approaches them. They are afraid of humans because their habitual experiences with them are negative, rather than neutral or positive; aversive conditioning teaches the animals fear of personnel.

photo 54

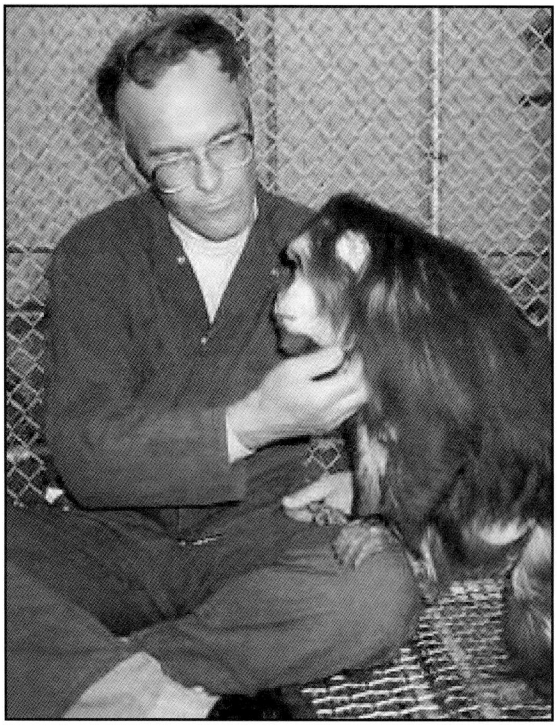

photo 55

It is often overlooked that nonhuman primates are sensitive social animals who respond to friendly attention from caretakers and investigators by gradually overcoming their conditioned fear and anxiety and establishing affectionate relationships with them.

photo 56

Positive interaction with monkeys and apes is essential for the well-being of the animals, data validity, and ease of handling (Wolfle, 1987). *The bond with the caregiver conveys to an animal a quiet sense of assurance upon which coping strategies can be developed* (Wolfle, 1996).

photo 57

photo 58

The behaviour of an animal during a procedure depends on the confidence it has in its handler. This confidence is developed through regular human contact and, once established, should be preserved (Home Office, 1989).

Positive relationships that develop between facility personnel and laboratory animals may result in an overall reduction in stress for the animals and may serve to buffer the potential stress of certain experimental situations resulting from the novelty of the procedure area, disease conditions, or certain experimental procedures (Bayne, 2002).

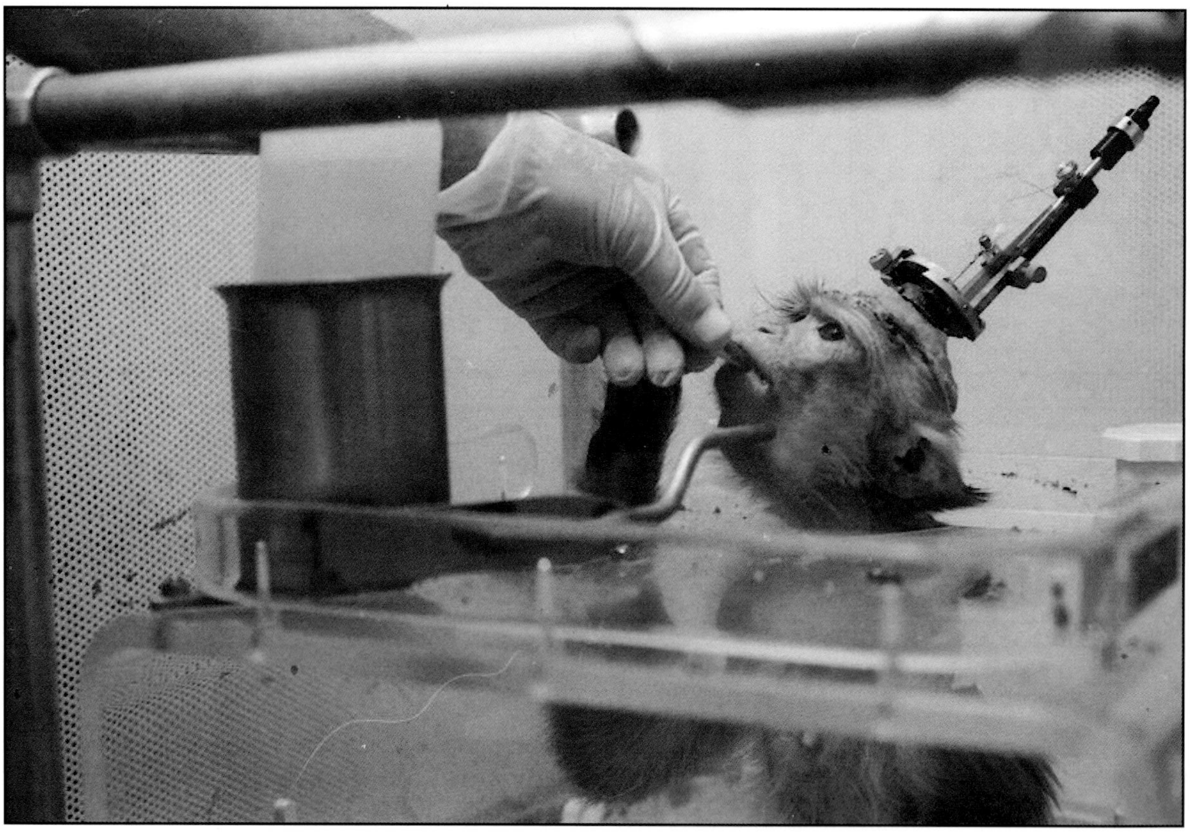

photo 59

Almost every animal commonly used in the laboratory responds positively to a little tender loving care. It's inexpensive, readily portable, safe even at the highest doses and spreads rapidly through the staff (Bennett, 1990). *The bond between people and animals in the laboratory, if understood and used consistently, can minimize certain variables related to stress in the animals* (American Association for Laboratory Animal Science, 2001). *Researchers must continue to question the barriers that have traditionally been erected against forming human-animal bonds in the name of objectivity and to investigate seriously the ways in which fostering the formation of such bonds can promote animal welfare without compromising the scientific respectability of research* (Russow, 2002).

Stroking and handling by humans can be a practical and effective technique for calming animals in situations where they are distressed, particularly animals that have been positively socialized by humans (Institute for Laboratory Animal Research, 1992).

photo 60

Animals who have developed a relationship based on mutual trust with attending personnel give the impression that they like human contact. This suggests that human contact has a relaxing, tension-releasing effect on them. Gantt et al. (1966) reports of a female rhesus macaque who was petted by a person for unspecified time periods 10 times on two different days. *On both days, significant decreases were found in heart rate during petting.*

Koban et al. (2005) exposed four male long-tailed macaques to daily 10-minute positive reinforcement training sessions, using assorted foods as a reward, for two months; four control subjects received no training sessions. *Results indicated that there was a statistically significant reduction in cortisol for trained subjects; cortisol in control subjects did not decrease from week one to week eight. Heart rate for the duration of the study proved to be lower in trained versus control subjects.*

Baker (2004) increased the time caretakers spent visiting (playing with, grooming, treat- feeding, talking to) seven adult female and five adult male chimpanzees housed in pairs and trios from two to four hours. Behavioral data were collected systematically, not during the visits, but between them. Therefore, the carry-over effect of human interaction, not the behavior during visits, was assessed. When the daily time of unstructured affiliation with personnel was doubled, the chimpanzees seemed to be more relaxed; they spent more time grooming each other (level of significance $p<0.05$) and less time engaged in agonistic displays (level of significance $p<0.06$).

A positive human-animal relationship based on mutual trust and respect is the basic condition to obtain the cooperation of nonhuman primates during procedures that would otherwise require involuntary restraint and incur distress for the animal and risks for the human handler.

3.1.6. Positive Reinforcement Training

Positive reinforcement training achieves two goals at the same time:

1. Intellectual stimulation for the animal subject and for the human caregiver (*Environmental Enrichment*);
2. Reduction of distress reactions of the animal subject and increase in safety of the human caregiver during husbandry and research-related procedures (*Refinement*).

3.1.6.1. Injection

Successful training protocols to obtain the subject's cooperation during injection have been described for:

photo 61

- a single-caged adult male baboon by Levison (1994); approximately nine one-hour training sessions were required to achieve the goal of the training;

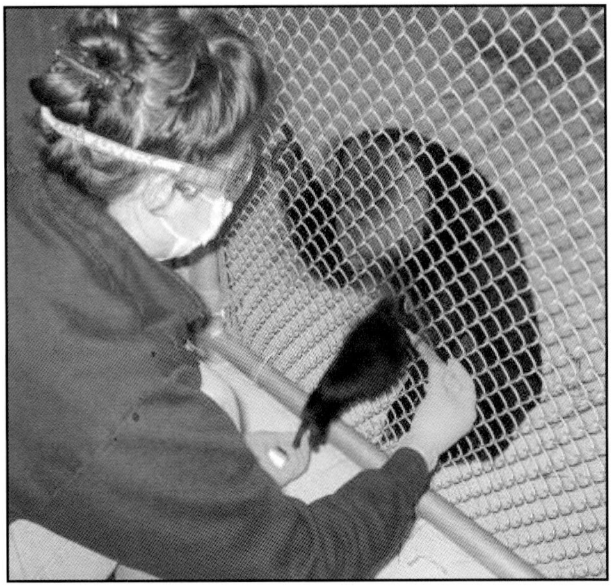

photo 62

- group-housed adult male lion-tailed macaques by Bayrakci (2003); the animals cooperated after a cumulative total of 1.5 to 5 hours of training.

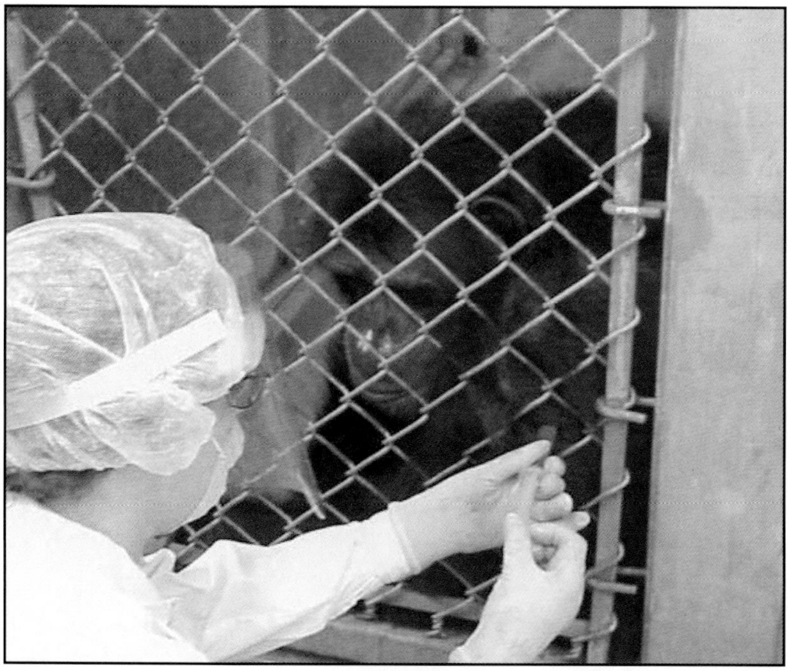

photo 63

Successful training protocols to obtain the subject's cooperation during injection have also been described for:

- a single-caged adult male mandrill (*Mandrillus leucophaeus*) by Priest (1991);
- a single-caged adult male mustached guenon (*Cercopithecus cephus cephus*) by Stringfield and McNary (1998);
- single-caged, pair- and group-housed female and male chimpanzees of all age classes by Spragg (1940), Schapiro et al. (2005), Videan et al. (2005a) and Russell et al. (2006); cumulative time investment to achieve cooperation during injection was about 87 minutes (Schapiro et al., 2005);

photo 64

- single-caged adult male rhesus macaques (LAREF, 2007c).

Cooperative behavior is consistently rewarded.

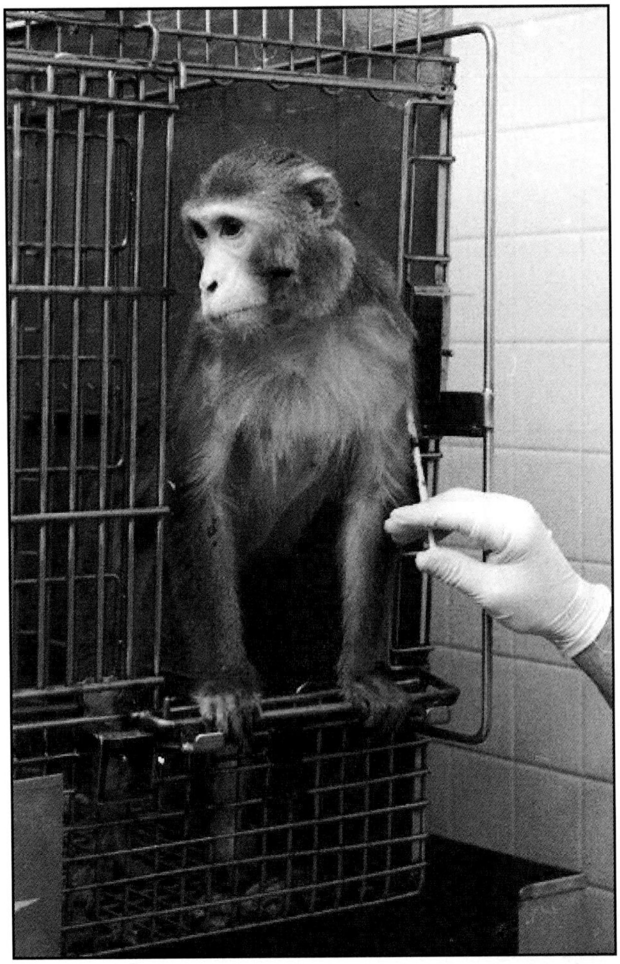

photo 65a

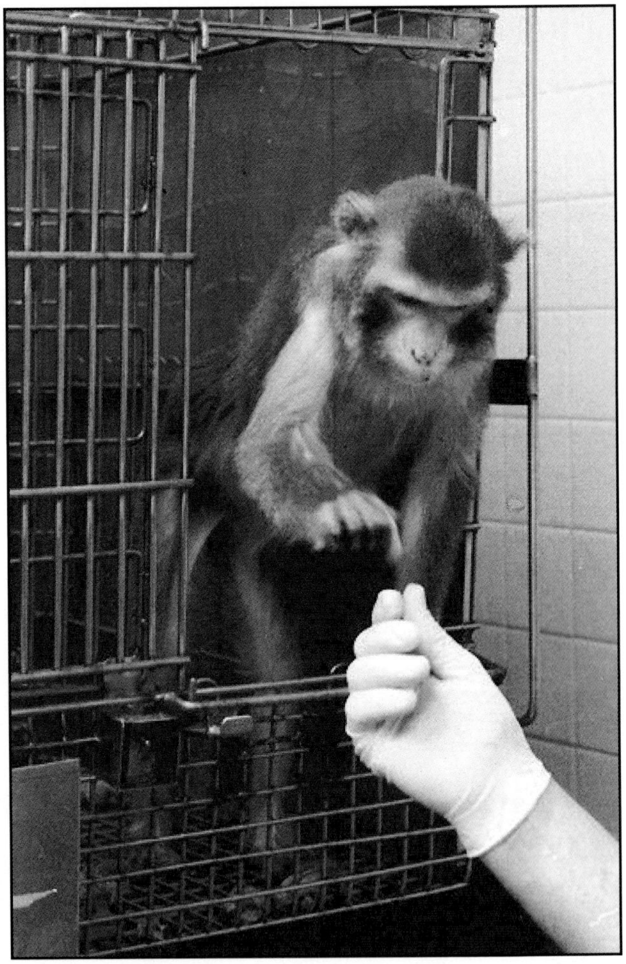

photo 65b

Bentson et al. (2003) compared the stress response to injection in four single-caged rhesus macaques who did not cooperate with the stress response of 17 single-caged rhesus macaques who had been trained to cooperate during injection. While serum cortisol concentrations did not increase in the trained subjects, cortisol increased significantly in the untrained subjects.

3.1.6.2. Blood Collection

Successful training protocols to obtain the subject's cooperation during venipuncture and subsequent blood collection have been described for:

- a single-caged adult male mandrill by Priest (1990);
- pair-housed adult female stump-tailed macaques by Reinhardt and Cowley (1992);
- pair- and group-housed chimpanzees of both sexes and all age classes by Laule et al. (1996), Schapiro (2000) and Schapiro (2005);
- single-caged and pair-housed rhesus macaques of both sexes and all age classes by Elvidge et al. (1976), Vertein and Reinhardt (1989), Reinhardt (1991b), and Phillippi-Falkenstein and Clarke (1992).

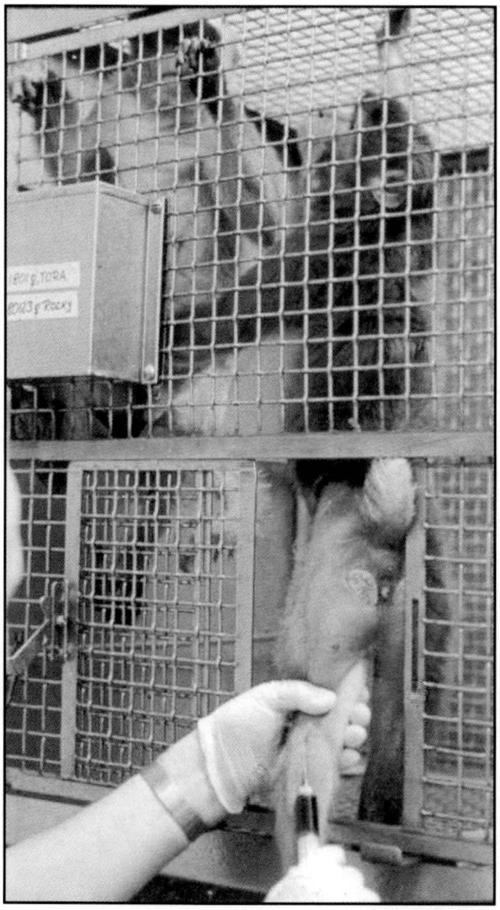

photo 67

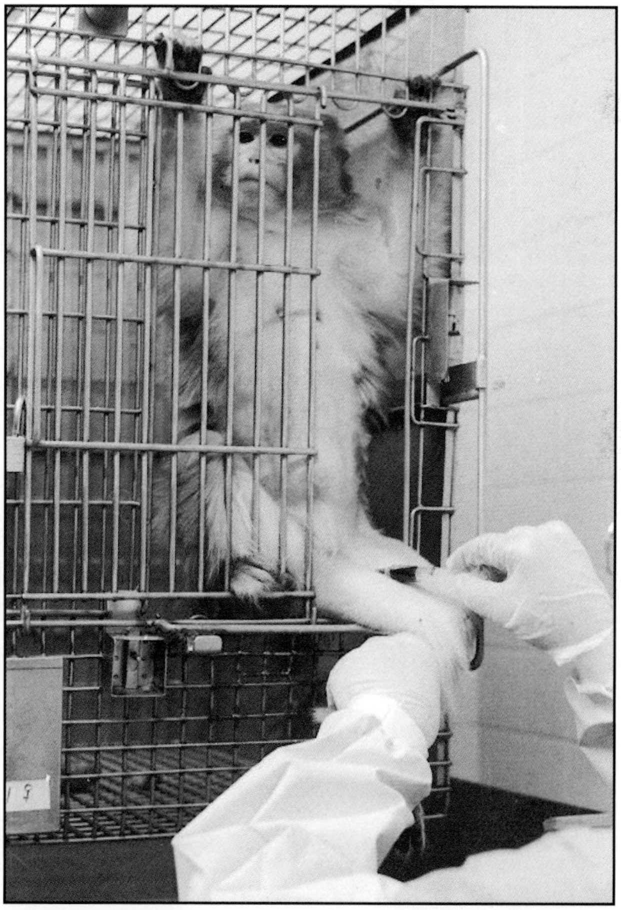

photo 68

Trained animals show no behavioral and no physiological stress response—as measured in changes in serum cortisol concentration (Elvidge et al., 1976; Reinhardt, 1991b; Bentson et al., 2003)—when they cooperate during blood collection.

photo 69

photo 70

Depending on the training technique applied, a cumulative mean total of 40 to 156 minutes are required to train adult male rhesus macaques to voluntarily present a leg or an arm for venipuncture in the home cage and hold still during subsequent blood collection (Reinhardt, 1991b; Pranger et al., 2006).

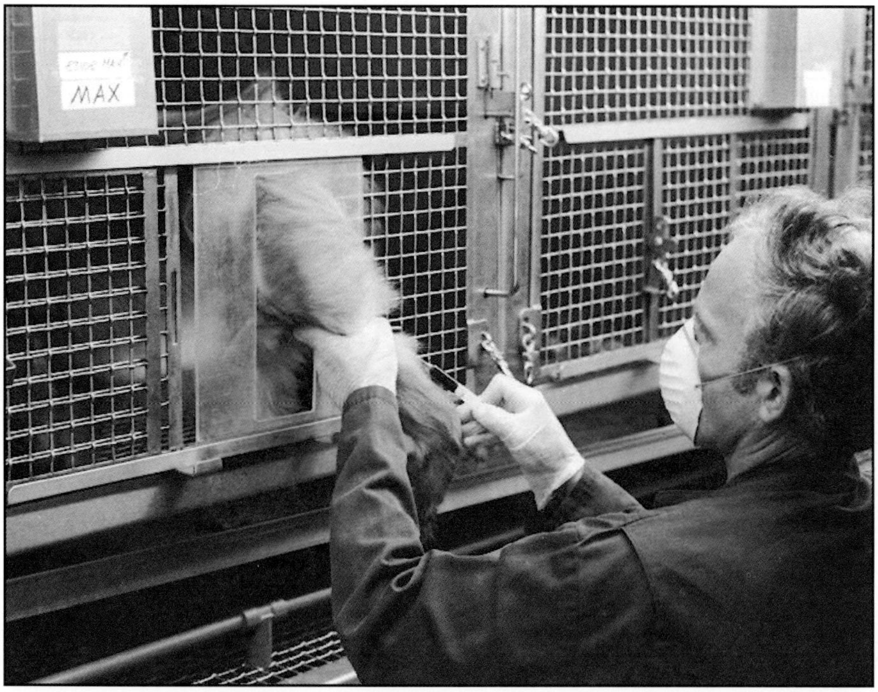

photo 71

Cooperation is always rewarded with praise and a food treat.

It has been argued that *monkeys can be trained to offer their arms or legs for blood collection with positive reinforcement, but this requires a considerable amount of time and dedicated staff* (Hrapkiewicz et al., 1998). It is true that dedicated staff is needed to establish and foster a trustful relationship with the animals in order to create a safe work environment for the training. The scientific literature, however, indicates that the time investment does not have to be "considerable" and, hence, should not be accepted as an excuse *not* to implement a positive reinforcement training program.

photo 72

photo 73

photo 74

photo 75

photo 76

Less than a cumulative total of 60 minutes of training time is needed to assure that adult female stump-tailed macaques cooperate during blood collection in the home cage and show no cortisol increase during this procedure (Reinhardt and Cowley, 1992).

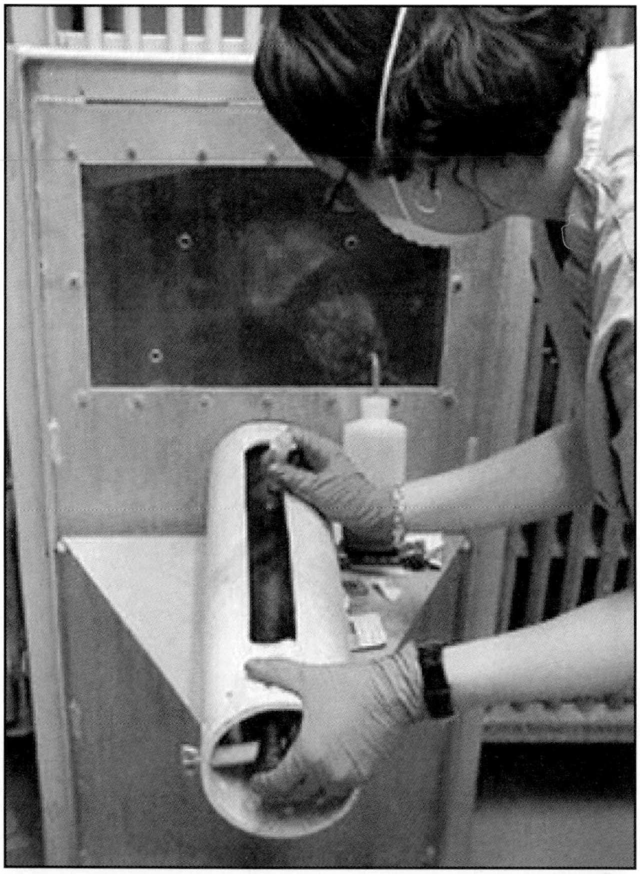

photo 77

It takes a cumulative total of about 3.5 hours to train chimpanzees to voluntarily hold on to a rod at the end of a "blood sleeve" and keep still during venipuncture and blood collection.

photo 78

3.1.6.3. Blood Pressure Measurement

Successful training protocols to obtain the subject's cooperation during blood pressure measurement have been described for:

- group-housed adult female and male woolly monkeys (*Lagothrix lagotricha*) by Logsdon (1995);
- single-caged adult male baboons by Mitchell et al. (1980) and Turrkan et al. (1989).

3.1.6.4. Urine Collection

Successful training protocols to obtain the subject's cooperation during urine collection have been described for:

- group-housed adult male vervet monkeys by Kelly and Bramblett (1981);
- group-housed adult female white-faced sakis (*Pithecia pithecia*) by Shideler et al. (1994);
- single-caged and group-housed juvenile and adult chimpanzees by Laule et al. (1996) and Lambeth et al. (2000);
- group-housed juvenile and adult marmosets of both sexes by Anzenberger and Gossweiler (1993), McKinley et al. (2003) and Smith et al. (2004);
- group-housed adult female tamarins (*Leontopithecus rosalia*, and *Saguinus imperator*) by Snowdon et al. (1985) and Smith et al. (2004).

3.1.6.5. Vaginal Swabbing

A successful training protocol to obtain the subject's cooperation during vaginal swabbing has been described for group-housed stump-tailed macaques by Bunyak et al. (1982). By the end of five training sessions of unspecified duration *it was no longer necessary to net and restrain the females. Indeed, some of them began to voluntarily approach the researcher and present for vaginal swabbing. Other females had to be cornered and gently contacted on the hips before they would accept a swab, often while holding the cage wire.*

3.1.6.6. Semen Collection

Successful training protocols to obtain the subject's cooperation during semen collection have been described for:

- group-housed gorillas by Brown and Loskutoff (1998);
- group-housed chimpanzees by Perlman et al. (2003).

3.1.6.7. Oral Drug Administration

Successful training protocols to obtain the subject's cooperation during oral drug administration have been described for:

- group-housed adult cotton-top tamarins (*Saguinus oedipus*) of both sexes by Savastano et al. (2003);
- single-caged adult male baboons by Turrkan et al. (1989);
- single-caged and group-housed adult marmosets of both sexes by Peterson et al. (1988) and Donnelly et al. (2007);

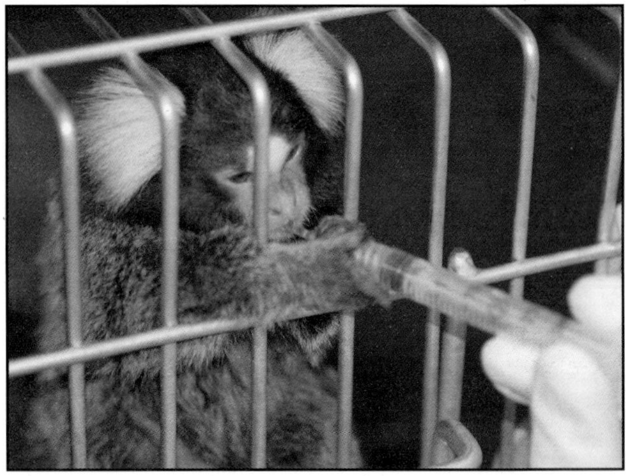

photo 79

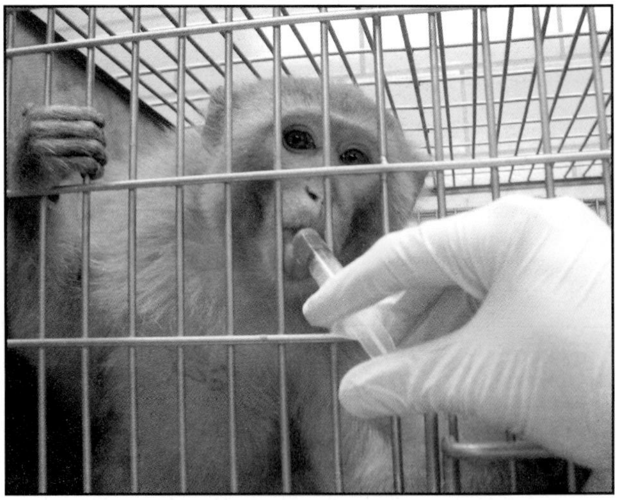

photo 80

- single-caged adult male rhesus macaques by Winterborn (2007).

3.1.6.8. Saliva Collection

Successful training protocols to obtain the subject's cooperation during saliva collection have been described for:
single-caged adult male rhesus macaques by Lutz et al. (2000);
- single-caged adult male squirrel monkeys by Tiefenbacher et al. (2003);
- single-caged and group-housed adult marmosets of both sexes by Cross et al. (2004).

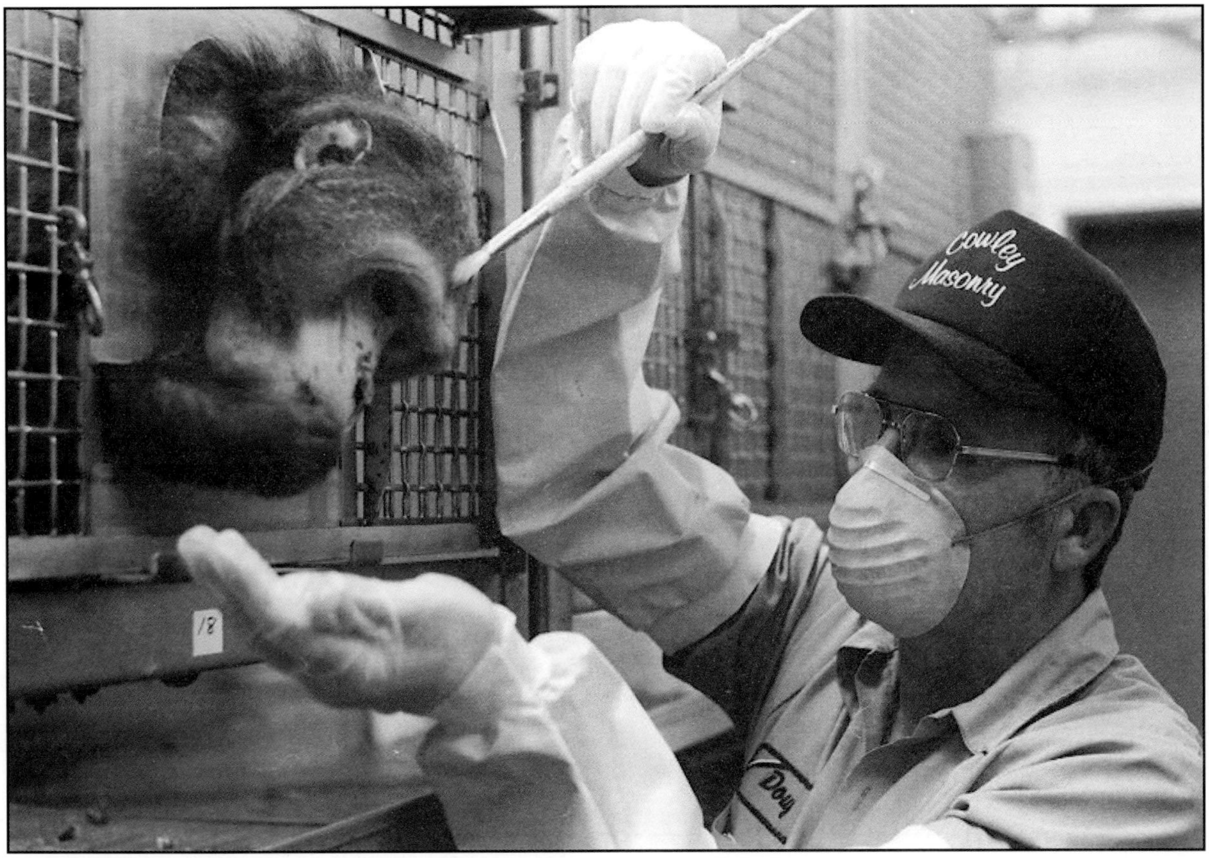

photo 81

3.1.6.9. Topical Treatment

Successful training protocols to obtain the subject's cooperation during topical treatment have been described for:
- group-housed adult female gorillas by Segerson and Laule (1995);
- group-housed female and male chimpanzees of all age classes by Perlman et al. (2001); and
- pair-housed adult stump-tailed macaques of both sexes by Reinhardt and Cowley (1990).

3.1.6.10. Weighing

A successful training protocol to obtain the subject's cooperation to climb onto scales for weighing has been described for pair-housed adult marmosets by McKinley et al. (2003) who invested a cumulative total of about one hour per pair to achieve the goal of the training.

photo 82

3.1.6.11. Pole Attachment and Chairing

Successful training protocols to obtain the subject's cooperation to allow having a pole attached to a permanent neck collar and being led to and securely placed in a restraint chair have been described for:

- single-caged male pig-tailed macaques of unspecified age by Nahon (1968);
- single-caged adult long-tailed macaques of both sexes by Skoumbourdis (2008);
- single-caged juvenile and adult rhesus macaques of both sexes by Skoumbourdis (2008).

photo 83

3.1.6.12. Capture

Successful training protocols to obtain the subjects' cooperation to move to a holding area or exit into a transfer cage have been described for:

- groups of bonobos (Pan paniscus) by Bell (1995);
- groups of Japanese macaques by Goodwin (1997);
- groups of chimpanzees by Kessel-Davenport and Gutierrez (1994) and Boomsmith et al. (1998).
- groups of rhesus macaques by Reinhardt (1990c). In order to train a heterosexual group of 45 rhesus macaques to voluntarily cooperate during the routine one-by-one capture procedure, an average of 20 minutes was invested per group member and 15 hours, respectively, for the whole group. It took about 15 minutes to catch all 45 animals without distressing them (Luttrell et al., 1994).

photo 84

3.2. Feeding Enrichment

Feeding enrichment promotes non-injurious food searching, food retrieving and/or food processing behavior.

photo 85

Nonhuman primates—here a troop of baboons—are biologically programmed to spend a major portion of their time searching for, retrieving and processing food. Therefore, food should be presented in the captive environment in such a way that they can engage in some kind of foraging behavior.

photo 86

3.2.1. Vegetables and Fruits

Attending care personnel typically work under time pressure. To have them chop supplemental vegetables and fruits for the animals in their charge is a waste of time. The animals have all the time needed to process the material themselves, and they like to do it.

photo 87

Offering whole, rather than already processed, vegetables and fruits of the season provides effective feeding enrichment without extra time investment. It introduces variety into the monotonous standard feeding regimen of commercial pelleted dry food and allows the animals to engage in species-typical food processing behaviors. Every animal should receive at least one medium-size whole fruit or vegetable on a daily basis. Animals living in pairs or groups should always have access to two fruits or vegetables to avoid competition and possible conflicts.

photo 88

photo 89

The following unprocessed produce has been fed to captive primates without any adverse side effects: apples, oranges, bananas, grapes, watermelons, pumpkins, squash, potatoes, carrots, string beans, corn on the cob, lettuce, celery, artichokes, bell peppers, sugar cane, cranberries, raspberries, coconuts, and peanuts in the shell (Bloomsmith et al., 1988; Spector and Bennett, 1988; Hayes, 1990; Beirise and Reinhardt, 1992; Nadler et al., 1992; Williams et al., 1992; Logsdon , 1994; Waugh, 2002; LAREF, 2007d). When presented behind a barrier—for example behind the bars or mesh of the enclosure—whole fruits and vegetables promote not only food processing, but also skillful food retrieval behavior.

Beirise and Reinhardt (1992) distributed every week 1 kg whole peanuts and on a different day 32 ears of corn to a 16-member breeding group of rhesus macaques. After a habituation period of eight weeks, 2-hour observations were conducted immediately after peanuts or corn were distributed in weeks 9, 10 and 11. Individual animals spent about:

- 77 percent of time husking corn ears, chewing husks and eating corn kernels and
- 47 percent of the time cracking peanut shells and eating peanuts.

photo 90

photo 91

I give whole corn with the husk to our pair- and group-housed rhesus and baboons. They love it, and I enjoy observing them "peel and eat," leaving a big mess after they have finished. They gnaw the cob into little pieces that finally fall through the grid floor on the pans. I don't mind cleaning up the mess; it's worth the treat (LAREF, 2007d).

3.2.2. Standard Food Ration Behind a Barrier or on Woodchips

Offering the daily food ration not freely accessible on the floor or in standard food boxes, but behind the bars or mesh wall/ceiling of the enclosure, is probably the easiest way of increasing the time that the animals spend obtaining and processing their food.

Reinhardt (1993a) distributed the daily biscuit ration of eight adult, pair-housed rhesus macaques first in their ordinary freely accessible food boxes, and then for a 2-week period, he threw the ration on the 22 x 22 mm-mesh ceiling of the cages. Time spent retrieving biscuits was recorded for each animal during four hours following food distribution.

- When the ration consisted of 66 small, bar-shaped biscuits, average foraging time increased 80-fold, from 17 seconds to 1363 seconds.
- When the ration consisted of 32 large, star-shaped biscuits, average foraging time increased 296-fold, from 12 seconds to 3551 seconds.

Working for their daily biscuit ration did not affect the males' body-weight balances.

photo 92

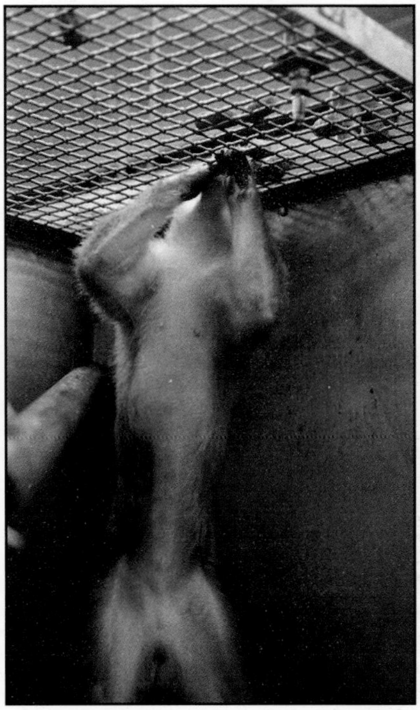

photo 93

photo 94

Reinhardt (1993b,c) observed eight pair-housed, adult male rhesus macaques and five female and seven male single-caged adult stump-tailed macaques, each for 30 minutes after their daily biscuit rations were distributed either in the ordinary food boxes with 73 x 47 mm access holes or in the same boxes remounted onto the 22 x 22 mm-mesh front panels of the cages a few centimeters away from the original access holes. All animals were habituated during 30 days to receiving their food in the *food puzzles*; their body weights did not change in the course of that time period.

- Rhesus macaques spent, on average, less than 1 percent of the time collecting biscuits from the food box versus 61 percent of the time retrieving them from the food puzzle.
- Stump-tailed macaques also spent, on average, less than 1 percent of the time collecting biscuits from the food box, but 63 percent of the time retrieving them from the food puzzle.

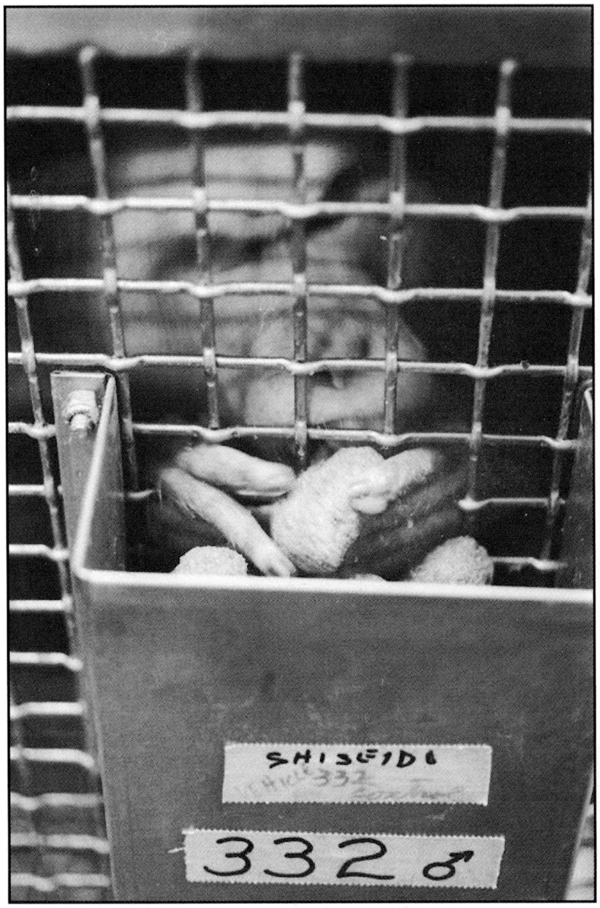

photo 95

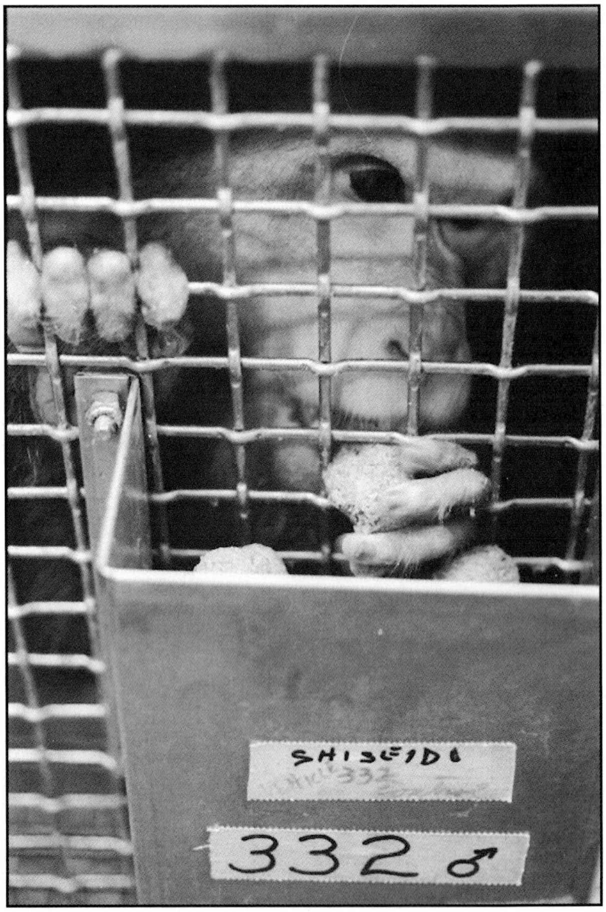

photo 96

Bertrand et al. (1999) report of four single-caged rhesus macaques, of unspecified age and gender, who received their daily pellet ration in a freely accessible standard feeder, and four other single-caged subjects who received their pellet ration on four days in a foraging device fitted on the front of the cage. Manipulative skills were required to retrieve the pellets from this device. *Over 90 percent of the food was eaten within the first 15 minutes with the standard feeder, whereas it took 60 minutes to reach this percentage using the foraging feeder. The amount of waste food was up to 17 times lower* when the animals had to work for their food instead of collecting it freely.

photo 97

Bloom and Cook (1989) mounted a commercial *puzzle feeder* on the front panel of the cages of two adult male rhesus macaques and habituated the animals to retrieving their daily single portion of biscuits from the device. It took the two males 20 to 30 minutes to retrieve their food.

Murchison (1994) distributed the daily biscuit ration of 16 single-caged pig-tailed macaques of both sexes and various age classes for a four-day period in the standard feeder or in a custom-made *forage feeder*. During the first hour after biscuit distribution, *the animals spent significantly more* [unspecified] *time foraging with the forage feeder than the standard feeder.* They *consumed nearly all the food received from the forage feeder, leaving less on the cage floor to become contaminated.*

Murchison (1995) videotaped the behavior of 20 single-caged adult female pig-tailed macaques, each for one hour, when the ration of 40 biscuits was presented in the standard feeder with one big access hole (5 cm diameter) versus a same size feeder with four small access holes (3 cm diameter). The animals spent, on average, 11 percent of the observation hour using hands, teeth and feet to remove biscuits from the feeder with small holes versus only 1 percent of the time to collect biscuits from the standard feeder with one big access hole; the difference was statistically significant. Unlike with the standard feeder, the animals consumed most of the biscuits they retrieved from the test feeder; this implied that fewer pieces of biscuits were dropped on the floor.

Beckley and Novak (1989) mounted foraging racks high up on the front of the enclosures of three groups of 3 to 6 rhesus macaques of different age classes and both sexes and compared their feeding behavior when the standard pellet ration was distributed in these racks, as opposed to the traditional practice when the ration was spread on the floor. The animals were tested daily in each condition until all food was eaten for a period of three weeks.

When they had to climb up to the racks, reach through the mesh and retrieve pellets, they were in contact with pellets for a significantly longer time than when the food was available freely accessible on the floor.

Lutz and Novak (1995) compared the behavior of three heterogeneous groups of four rhesus macaques during one-hour observation periods, after their daily biscuit ration was thrown on the bare floor versus when the biscuits ration was mixed with wood shavings.

When the animals were required to forage through the shavings, they engaged in significantly fewer agonistic interactions than when their food was freely available and could be hoarded in piles and monopolized by dominant group members.

3.2.3. Expanded Feeding Schedule

Taylor et al. (1997) expanded the feeding schedule of a group of four adult female and one adult male bonnet macaques by portioning the daily ration of 150 biscuits and 1 cup of sunflower seeds and dispersing one half of the ration on the woodchip litter at the usual time in the morning, and the other half in the afternoon. In the course of 10 weeks, the animals were observed during several 10-minute sessions starting one hour after food distribution.

When they received their daily food ration in two small portions (week 6-10), rather than in one big portion (week 1-5), they spent twice as much time foraging (about 52 versus 26 percent), probably because it was more difficult for them to find the food in the woodchips.

photo 98

3.2.4. Special Food in and on Gadgets

Numerous gadgets baited with special food treats—rather than the standard food—have been developed to encourage foraging-related activities in captive primates.

Bayne et al. (1992) secured Plexiglas boards covered with artificial turf inside the single- cages of eight adult male rhesus macaques. Commercial, flavored food particles were sprinkled on the turf boards daily two hours after the morning feeding; this was followed by 30-minute observations of each subject on 20 days over the course of a six-month period.

The males *foraged*, on average, 52 percent of the observation sessions; there were no signs that the they lost interest in foraging from the turf boards over time.

Riviello (1995) scattered wheat seeds on the two turf boards positioned in the cages of two small groups of capuchin monkeys, of unspecified gender and age, and observed the animals 16 times during the first 30 minutes. Individuals foraged from the boards 6 to 76 percent of the observation time.

Fekete et al. (2000) mounted a turf board inside, on a shelf of the cages of 10 pair-housed adult female squirrel monkeys and sprinkled a mixture of nuts, seeds and dried fruits onto the board on 11 consecutive days, right after the normal food was distributed. During the first 20 minutes, individuals spent approximately 36 percent of the time foraging.

Lutz and Farrow (1996) mounted turf boards to the outside of the front panel of the cages of 10 adult female long-tailed macaques and sprinkled sunflower seeds on the turf every morning, after the animals had received their daily biscuit ration. During three weekly 30-minute observations conducted at random times over a period of eight weeks, the animals spent an average of 11 percent of the time foraging. The boards were used by the animals with consistency; there was no indication that they lost interest in them over time.

Lam et al. (1991) tested six single-caged adult male long-tailed macaques on six days during the first hour after the animals received a fleece cushion sprinkled with commercial tidbits. In order to enhance the animals' interest in the supplemental treats, the daily food ration was withheld until after completion of the tests. The males spent, on average, 8 percent of the observation hour picking out food crumbles from the fleece.

Spector et al. (1994) furnished the drop pans of 24 single-caged baboons of unspecified age and gender with *foraging trays*. Every other afternoon, a mixture of seeds, dried fruits, pieces of vegetables, alfalfa cubes, feed corn and dog biscuits was added to the tray and then covered with a thin layer of fresh hay. The baboons had to reach through the bars of the cage floor, search for food items and then retrieve them. The animals were not systematically observed, but a review of many hours of video recordings taken during two years indicates that the animals spent 30 to 120 minutes per day foraging.

Chamove and Scott (2005) made 6-hour video recordings of four family groups (five to 11 individuals) of cotton-top tamarins on two consecutive days when they were presented with a *forage box* to which they were extensively habituated. The box was filled with a mixture of sawdust and small food items. Over the six hours, any given monkey was engaged in searching for and retrieving food from the box approximately 7 percent of the time.

Molzen and French (1989) suspended a plastic probe feeder, filled with broken corn cob and raisins and closed with an opaque lid with 3-cm-diameter access hole, in the enclosures of three golden tamarin families. During 5-minute test sessions conducted on seven days, engagement in extractive foraging was, on average, 5 percent of the time for adults and 27 percent of the time for juveniles.

Bryant et al. (1988) released six individually caged, adult male long-tailed macaques, one animal at a time, for 30 minutes into a playpen on 12 days, distributed over a three-

photo 99

week period. The playpen was furnished with a nylon ball, a telephone directory, a nylon rope and a tray placed below the grid floor of the cage, containing woodchips scattered with sunflower seeds and peanuts. The animals showed little interest in the enrichment items, but spent about 33 percent of the time reaching through the wire mesh of the cage floor and retrieve seeds and peanuts.

Hayes (1990) placed the daily feed, along with fruit treats and nuts, in two custom-made probe feeders consisting of 178 mm-diameter PVC pipes that were divided into three sections and had access holes of different sizes. The center section of the pipes was filled with hay into which part of the food was mixed. The two feeders were hung in the enclosure of a group of two adult female, one adult male, one juvenile and one infant capuchin monkeys. During five 60-minute observations conducted right after the feeders were filled, the animals spent, on average, 30 percent of the time selecting food, gathering food, processing and consuming food.

Bloom and Cook (1989) loaded a commercial *puzzle feeder* daily with 10 peanuts in the shell and tested two single-caged adult male rhesus macaques. The total time spent retrieving and consuming the peanuts ranged between 10 and 15 minutes.

Brent and Long (1995) made use of a perforated PVC *pipe feeder* filled with a mixture of peanut butter, marshmallows, corn, sunflower seeds and macaroni. The gadget was attached outside to the cage panels of four adult female, single baboons on two consecutive days. During subsequent one-hour observations, the animals spent, on average, 85 percent of the time retrieving food.

Prist et al. (2008) replenished three suspended *feeder balls,* made of woven vines, regularly with leaves, straw and food treats and recorded the behavior of a group of one adult female, one adult male and two juvenile howler monkeys (*Alouatta guariba*) for a about 60 hours. Subjects spend approximately 11 percent of the observation time contacting the balls.

photo 100

Bjone et al. (2006) tested four pairs of adult female marmosets who were accustomed to a perforated feeder box that required the animals to swing discs over holes in order to uncover and retrieve food rewards. During 20-minute observation sessions individual animals spend about 35 percent of the time with the feeders.

Steen (1995) placed two different feeders, each for a period of two weeks, into the enclosure of two adult male and one adult female cotton-top tamarins and observed the animals during one-hour sessions. No observations were made on the first day of feeder presentation to exclude novelty effects in the behavioral data.

- When the animals had access to three PVC boxes with tubular access holes on either side and filled with *primate cake*, they spent about 23 percent of the time retrieving food.
- When they had access to a perforated bamboo pipe filled with bran and mealworms, they spent about 22 percent of the time retrieving food.

McGrew et al. (1986) designed a *gum-tree* consisting of seven stacked cylindrical blocks, each containing four reservoirs filled with gum arabic. To gain access to a reservoir, a monkey had to gnaw through 1 to 3 mm of wood. The gum tree was tested in 10 female and 23 male marmosets who lived in three family groups. Each family was observed for the first 30 minutes after a filled gum-tree was fixed vertically in the cage once every day for a one-week study period.

Individual animals spent, on average, 51 percent of the observation time in contact with the gadget showing the full range of species-typical gum-foraging patterns, including urine-marking of breached reservoirs.

photo 101

Roberts et al. (1999) injected acacia gum into 2 to 3 cm deep holes of 30 cm long branch segments and placed one *gum feeder* each in the cages of 28 adult marmosets of both sexes, living alone (n=16) or in pairs (n=12). The feeders were left in the cages for five days and the animals tested for 30 minutes right after gum was injected into the branch on day 1, day 3, and again on day 5.

The marmosets spent, on average, 43 percent of the observation time gum-foraging on day 1, and 10 percent of the time on day 5. The branches were already heavily gouged on day 5.

photo 102

photo 103

Maki et al. (1989) designed metal *pipe feeder puzzle boxes* containing sticky foods—such as applesauce, mashed bananas, spaghetti sauce, and dry fruit drink powder. Four adult chimpanzees, living with other companions in pairs or trios, were observed during approximately eight 30-minute sessions distributed over a period of one month, when a regularly filled pipe feeder was permanently mounted from outside on the chain link fencing of the home quarters. The four subjects spent, on average, 23 percent of the time manufacturing dipping sticks from branches, and an additional 30 percent of the time fishing with these tools for the moist foodstuff in the box.

Celli et al. (2003) mounted an open transparent polyethylene bottle, which was filled daily with honey, in front of the cages of three pairs of adult female chimpanzees and offered them plastic brushes, wires, chopsticks and rubber tubes from which they could chose suitable tools for retrieving honey from the bottle. During daily one-hour observations [probably right after presentation of the bottle], the animals spent about 9 percent of the time checking out suitable fishing tools, and 31 percent of the time retrieving honey.

Lambeth and Bloomsmith (1994) conducted six 30-minute observations of eight adult female and six adult male chimpanzees, living in pairs or groups of four, after a PVC pipe cut in half and planted with rye grass was attached to the front panel of the chain link fencing of the subjects' enclosures. The animals spent, on average, 4 percent of the time picking grass with their fingers through the fencing. When sunflower seeds were added to the grass on six additional occasions, individuals spent 20 percent of the time searching for and picking up seeds.

Brent and Eichberg (1991) attached one Plexiglas sheet with holes on the mesh ceilings of the enclosures of eight heterogeneous groups of three or four chimpanzees. After a 7-day habituation period, commercial food treats were placed on these *puzzle boards* on four different occasions and the animals' response was recorded during one-hour sessions. The chimpanzees manipulated the puzzles and consumed treats, on average, 17 percent of the observation time.

Gilloux et al. (1992) monitored a heterogeneous group of seven chimpanzees for 12 two-hour sessions when a 15 cm-diameter plastic pipe filled with fruits, vegetables and biscuits was attached outside onto the welded mesh of the enclosure. The apes could manipulate food items to the open end of the pipe by inserting bamboo canes or willow twigs through holes drilled along the side of the pipe facing them. Individuals *used* the filled feeder, on average, 18 percent of the observation periods.

3.2.5. Special Food Mixed with a Substrate

Wood shavings in the catch pans provide an ideal substrate to foster foraging activities. On days when we change the pans—three times a week—we sprinkle sunflower seeds on the shavings. Our rhesus and squirrel monkeys then search with their fingers through the litter and pull the seeds through the floor grids, eat them or store them in their cheek pouches. Since we change the pans, rather than dump the bedding, we don't have any drainage problems in the rooms. This feeding enrichment technique doesn't require undue extra work time in our colony of approximately 130 monkeys. I'd say the benefit of being able to provide even a brief period of foraging behavior for our caged primates is worth the little additional time it takes to put the bedding in the pans and add a handful of seeds (LAREF, 2007d).

Bryant et al. (1988) tested six adult male long-tailed macaques alone in a relatively big cage each day for 30 minutes. The cage had a tray placed below the grid floor containing woodchips mixed with sunflower seeds and peanuts. Individuals spent approximately 37 percent of the time reaching through the grid floor, searching for and retrieving food from the woodchip litter. The interest in this activity increased over the course of a 12-day study period.

photo 104

Anderson and Chamove (1984) spread a mixture of grain in the morning and in the afternoon on the woodchip litter of a heterogeneous group of eight stump-tailed macaques on two consecutive days. During a cumulative total of 110 minutes of observations conducted on each of the two test days, individuals were seen foraging approximately 30 percent of the time.

Boccia (1989b) and Boccia and Hijazi (1998) observed a group of seven adult female, one adult male and seven juvenile pig-tailed macaques two weeks before and two weeks after sunflower seeds were scattered once every day on the woodchip litter. Individuals were observed in both conditions during two 5-minute sessions. They spent, on average, 15 percent of the time searching for seeds in the woodchips. During the control condition, they were engaged in partner-directed hair pulling on 26 occasions; when seeds were scattered on the woodchips and the animals more engrossed in foraging, hair-pulling was witnessed only three times. Similarly, the incidence of fighting was significantly lower during the seed-foraging condition than during the control condition.

Blois-Heulin and Jubin (2004) found in a family group of red-capped mangabeys (*Cercocebus torquatus torquatus*) that the animals searched for supplemental seeds significantly longer when the seeds were distributed on straw litter rather than on the bare ground. The increased foraging activity was paralleled by a significant decrease in self-directed behaviors.

Straw and woodchip litter are good foraging substrates also for other primates such as long-tailed and rhesus macaques.

photo 105

photo 106

Grief et al. (1992) studied a group of five adult chimpanzees and 17 single-caged adult chimpanzees of both sexes. Each subject was observed for three 15-minute sessions per day, three times a week. A bedding of straw or shredded paper mixed with sawdust was continuously available to all animals. Different foraging types were scattered on the bedding each morning at 8:00 for a period of one week, and random observations were carried out between 8:15 and 13:30. Subjects foraged about:

- 54 percent of the time for a mixture of Milo, cracked corn and wheat,
- 31 percent of the time for a mixture of rolled corn, barley and molasses, and
- 15 percent of the time for unsalted popcorn.

Baker (1997) provisioned seven adult female and six adult male pair- or trio-housed chimpanzees with straw, and scattered a mixture of sunflower seeds, peanuts, cracked corn, Milo and wheat twice a day over a period of nine weeks. Soiled straw was removed daily and replaced with fresh material every few days. Each subject was observed during 5-minute tests for a total of 10 hours.

On average, 5 percent of the time was spent searching for food and rearranging the straw, and an additional 4 percent throwing straw on oneself and cage mates, somersaulting and wriggling in straw, and using stalks to investigate otherwise out-of-reach features of their environment, including the keyholes of the locks on their caging.

3.3. Inanimate Enrichment

Inanimate enrichment increases the complexity of the living quarters and promotes non-injurious contact and interaction with objects.

3.3.1. Structural Enrichment

photo 107

All nonhuman primates—here a group of long-tailed macaques—spend the night and a great portion of the day on elevated sites at a safe distance from ground predators, so it is logical that their enclosures in research facilities can be distinguished as being primate-adequate *only* if they are furnished with high structures allowing the animals to retreat to and rest out of reach of the human predator. Such high resting surfaces are not really enriching the environment of the animals; they are a *necessity* and, therefore, should be a basic furniture for every living quarters of nonhuman primates.

Being permanently confined in the same enclosure is bound to foster social conflicts unless the partners have the option of breaking visual contact. For example, competition over food can be avoided when the animals are able to access food without being seen by dominant partners; aggressive intentions of a dominant animal—displayed as looking at, threatening or turning towards—can often be diffused when the subordinate target quickly disappears out of sight.

photo 108

photo 109

3.3.1.1. Structural Enrichment in Cages

The spatial limitation of the legally minimum-size standard cage can make it quite a challenge to open up the vertical dimension for the confined animal in a species-appropriate manner.

A high perch opens up the vertical dimension, thereby increasing the usable cage space and promoting species-adequate behaviors, such as climbing, leaping (if the cage is large enough), balancing, bouncing, perching, sleeping, looking-out, retreating to a safe place during alarming situations, and retreating to a dry place during the cage cleaning procedure. Access to a high resting site has survival value for nonhuman primates. This explains why they do not lose interest in high resting surfaces over time.

Schmidt et al. (1989) kept three subadult male rhesus macaques in single-cages, on unspecified locations on the cage rack, for 2 to 13 months. Each cage was equipped with a 72-cm long aluminum rod, 2 cm in diameter, mounted parallel to the sides and bottom of the cage at an unspecified height. During 10 half-hour observation sessions conducted at a time when the animals were not disturbed by personnel and noise, the animals were sitting on their perch, on average, 62 percent of the time.

Reinhardt (1989) assessed the time budgets of 25 adult male rhesus macaques who were housed in single-cages each equipped with a 120-cm long polyvinyl chloride (PVC) pipe that had a diameter of 5 cm and was installed diagonally, with a slope of 15°, about 40 cm above the floor. The males had been exposed to these perches for 12 months. There were 14 males in upper-row cages and 11 males in lower-row cages. During two hours of observations, when the animals were not disturbed in any manner, individual males sat on their perches 28 percent of the time. There was only one male (4 percent) who did not use his perch during the two hours. The average time spent on the perch was:

- 45 percent for the males in lower-row cages, versus
- 15 percent for the males in upper-row cages; the difference was statistically significant.

photo 110

The greater attractiveness of a perch for lower-row cage individuals was probably related to the fact that they lived closer to the ground and at a greater distance from the light source. Obviously, sitting on an elevated surface was more advantageous for them than for individuals in the high and bright upper-row cages.

photo 111

Woodbeck and Reinhardt (1991) confirmed these findings in 28 pairs of adult female rhesus macaques who lived in double cages, each furnished with two 12-cm long PVC pipes, located either in the bottom row (n=14 animals) or in the top row (n=14 animals). The females had been exposed to these perches for more than 24 months. During seven 30-minute observations conducted in the late afternoon when personnel were no longer in the building, average time spent perching was:

- 32 percent for the females in lower-row cages, versus
- 7 percent for the females in upper-row cages; the difference was statistically significant.

Similar findings were reported by Shimoji et al. (1993), who attached four parallel-connected PVC pipes, 5 cm in diameter, to the back of the cage 27 cm off the floor, of 10 female and 10 male adult, single long-tailed macaques for a three-day study period. Remote video recordings revealed that animals caged on the bottom row of the rack spent, on average, 26 percent of the day on the *perch,* while animals caged on the top row spent only 14 percent of the day on the perch; the difference was not statistically different, but it was consistent on each of the three days.

The usefulness of a perch depends on its placement in the cage. Primates are inquisitive animals, but they want to hide during alarming situations. Therefore, the perch should be installed in such a way that an animal can:

- sit right in front of the cage and check out what is going on in the room, and
- retreat to the back of the cage when being frightened, for example, when a person enters the room.

Bayne et al. (1992) exposed eight adult male rhesus macaques, each to a galvanized steel perch of unspecified diameter that was placed approximately 20 cm off the floor of the cage, parallel to the side wall. The animals were kept in single cages of unspecified location in the cage rack. During eight weekly 30-minute observation sessions, the animals sat on their perches, on average, 17 percent of the time without showing signs of habituation. All eight males used their perches.

Watson (1991) made three 20-minute observations during normal daytime working hours of 31 adult female and 31 adult male long-tailed macaques. The animals had lived for 14 weeks in upper- and lower-row single-cages, each equipped with three stainless steel rods with a 2 cm diameter running parallel to the back wall of the cage at a height of 18 cm. The animals perched on these rods, on average, 86 percent of the observation time [time budget of lower-row caged animals and upper-row caged animals was not compared].

photo 112

Boinski et al. (1994) observed one male and 15 female squirrel monkeys of unspecified age, who lived in cages equipped with one perch of unspecified diameter and height. Half of the animals lived in upper-row, and the other half in lower-row cages. Each animal was observed in the afternoon during four to eight 5-minute sessions distributed over a period of two weeks. Individuals sat on their perch about 87 percent of the time [time budget of lower-row caged animals and upper-row caged animals was not compared].

photo 113

In their natural habitats nonhuman primates have the tendency to avoid being on the ground but prefer to stay high up in trees or on other elevated places where they are safe from predators. It is not surprising that chimpanzees (Clarke et al., 1982; Goff et al., 1994; Ochiai and Matsuzawa, 2001; Ross and Lukas, 2006), squirrel monkeys (Salzen, 1989; Taylor and Owens, 2004), tamarins (Snowdon and Savage, 1989; Buchanan-Smith, 1991), rhesus macaques (Reinhardt, 1992a; O'Neill-Wagner, 1994; Kravic and McDonald, 2003; Clarence et al., 2006), bush babies (Günther, 1998; Watson et al., 2002), long-tailed macaques (Westlund, 2002), Japanese monkeys (Terazawa et al., 2002), marmosets (Buchanan-Smith et al., 2002) and presumably all other primate species try to occupy the highest resting surface of their enclosure.

If only one perch can be installed, it should be placed as high as possible, allowing an animal to (a) sit on it without touching the ceiling with the head and without touching the floor with the tail, and (b) use the space beneath the perch, if the floor space is insufficient, for free postural adjustments. The height of the cage should be determined by these conditions.

photo 114

When minimum-size standard cages are stacked in double tiers to accommodate a maximum number of animals in a room, these conditions are usually not met. There is insufficient height and perches are placed at a level that makes it impossible for an animal to turn around freely and adjust postures without touching the walls or stepping on the perch.

photo 115

photo 116

Dettling (1997) demonstrated in a choice test that marmosets prefer relatively thick perches over thin perches. The same probably holds true also for other primate species. A perch with a sufficiently large diameter, allowing an animal to sit firmly on it without holding on to the wall or ceiling, is likely to be more suitable, i.e., comfortable than a relatively thin rod on which the animal has to balance in order not to fall off.

Access to a perch allows pair-housed animals to quickly get away from each other when there is social tension; this can help reduce aggressive interactions. Kitchen and Martin (1995) observed five pairs of common marmosets, each for a total of 20 hours, when their cages were barren versus equipped with three perches, 2 to 3 cm in diameter. When they had access to perches, the marmosets stopped showing startle responses and the incidence of aggressive interaction was significantly reduced.

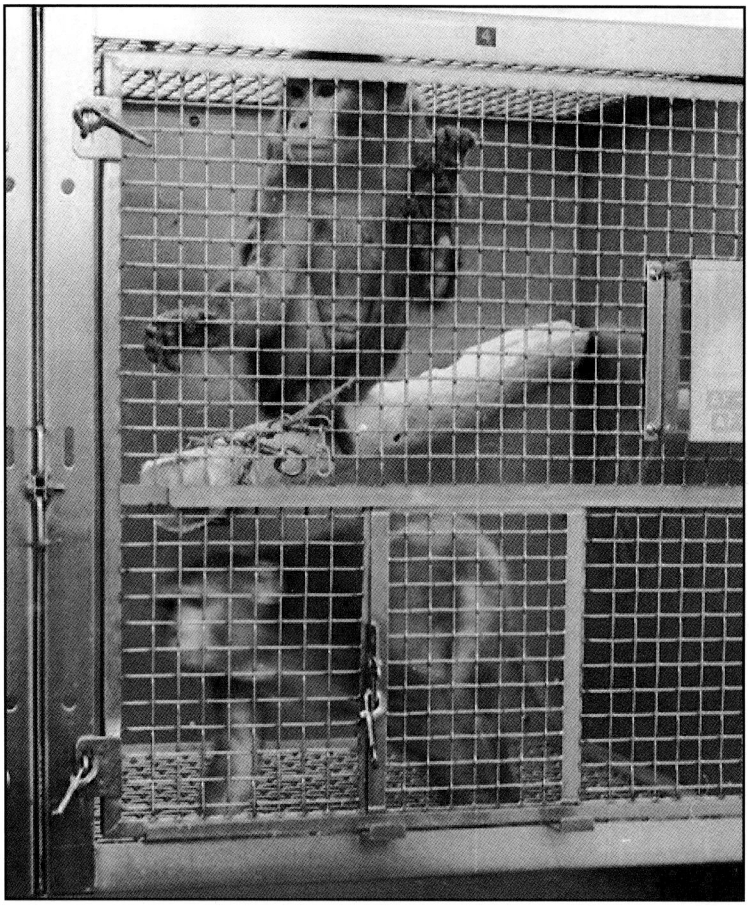

photo 117

In their natural habitat, primates very often leap from branch to branch, but they usually do not swing on thin branches or lianas. It is, therefore, not surprising that confined animals have little use for swings, especially since the small size of their living quarters does not provide sufficient space to actually swing back and forth.

Bryant et al. (1988) observed six single adult male long-tailed macaques daily for 30-minute sessions in a play cage that was equipped with a swing suspended 60 cm from the ceiling. During a test period of 12 days, two males never used the swing; the four others spent, on average, only 2 percent of the time on it.

Dexter and Bayne (1994) tested nine adult single-caged rhesus macaques of both genders in the presence of either two types of PVC swings, a hemp rope swing or a swing made of artificial vine. Each animal was exposed to the swings for a three-week period and observed three times for 30 minutes during this time. The animals manipulated the swings but showed little inclination to actually use them for swinging. Altogether swinging was witnessed only six times in the course of 360 minutes of observation, and the overall average percentage of time that a monkey was actually swinging was less than 1 percent.

Kopecky and Reinhardt (1991) installed a PVC perch in one section and a PVC swing at the same height in the other section of upper-row double cages of 14 adult, pair-housed rhesus macaques and observed each animal after one month for 60 minutes. Subjects spent, on average, 11 percent of the time on the perch, but only 1 percent of the time on the swing. It was concluded that the animals' statistically significant preference for the perch was probably related to the fact that the perch, unlike the swing, was a fixed structure permitting continuous relaxed postures rather than brief balancing. Moreover, the perch, unlike the swing, allowed the animals to sit right in front of the cage with visual control of the events going on in the room.

photo 118

The provision of small protruding verandas is probably the most species-appropriate, and at the same time practical way of offering caged primates the opportunity to get access to the arboreal dimension and retreat to a high, safe vantage point from which they have visual control of the surrounding environment.

photo 119

The spatial constraint of the cage makes it impossible to furnish it with structures in which an animal can take visual refuge from a dominant cage mate, but vertical blinds can readily be installed without occupying part of the floor area.

Basile et al. (2007) observed 18 male/male pairs, 2 female/female pairs and 5 male/female pairs for two 30-minute sessions before and one week after a *privacy divider* was placed in their double cages. The blind was oriented in such a way as to physically divide the front half of the cage, while leaving open access through the rear half. With the privacy divider in place, the animals spent significantly more time in the same half of the cage (52 versus 44 percent). It was concluded that *the increase in proximity associated with the presence of the privacy dividers reflects an increase in social tolerance and/or attraction. We suggest that the privacy divider may provide a safe haven and give monkeys the ability to diffuse hostile situations before they escalate.*

photo 120

photo 121

Reinhardt and Reinhardt (1991) inserted a *privacy panel*, consisting of a sheet of stainless steel with a rectangular 23 x 32 cm passage hole close to the back wall of the cage, between the two halves of each double cage of 15 adult female rhesus pairs. One-hour observations before and seven days after placement of the privacy panels revealed that companions:

- spent significantly more time in the same half of the cage (76 versus 61 percent),
- spent significantly more time engaged in affiliative interactions (37 versus 27 percent), and
- had fewer agonisitc disputes (0.3/h versus 2.2)

when they had the option of visual seclusion.

3.3.1.2. Structural Enrichment in Rooms and Pens

photo 122

photo 123

It is relatively easy to create species-adequate structural enrichment for primates who live in pens or rooms; there is usually sufficient vertical and horizontal space.

Elevated structures not only make the vertical dimension accessible to the animals, but they also provide them with easy ways of quickly getting away from each other in situations of potential social conflict.

Neveu and Deputte (1996) recorded the behavior of a breeding troop of gray-cheeked mangabeys (*Cercocebus albigena albigena*), consisting of three adult and two juvenile females and one adult and one subadult male, during 30-minute sessions when they lived in a barren cage versus a cage of the same dimensions but fitted with four perches at different heights. Access to perches decreased agonistic behaviors from about 25 to 0 percent of all interactions; at the same time it increased socially positive behaviors significantly from about 2 to 10 percent of all interactions.

Nakamichi and Asanuma (1998) tested a group of four adult female Japanese macaques in two identically sized enclosures that were either unstructured or furnished with eight wooden perches at different heights. Several 15-minute observation sessions showed that the average number of agonistic interactions was significantly lower in the furnished cage than in the unfurnished cage.

Caws et al. (2008) compared aggression-related injuries in a group of 15 adult female, five adult male and nine immature chimpanzees before and after a complex vertical structure was erected in the compound. The proportion of serious injuries out of the total recorded injuries was lower in the three years after than in the three years before the structure was erected (32 versus 59 percent).

Similar to the situation in cages, adult primates show hardly any interest in movable structures such as swings, ropes, suspended barrels and Ferris wheels, but they will spend most of the day and all night on fixed structures such as platforms, shelves, ladders, benches and perches well above the enclosure floor (langurs: Schwenk, 1992; rhesus macaques: Lehman and Lessnau, 1992; baboons: Kessel and Brent, 1996; chimpanzees: Howell et al., 1997).

photo 124

photo 125

photo 126

Movable structures, such as suspended branches and Ferris wheels, are more attractive for young animals than for adults.

photo 127

Visual barriers, which allow the animals to be out of sight of one another provide opportunities to avoid attacks and also prevent dominant individuals from restricting access of subordinates to other parts of the enclosure (Council of Europe, 2006).

Erwin et al. (1976) studied agonistic interactions between adult female pig-tailed macaques who lived in four breeding groups; there were approximately 12 females in each group. Observation sessions were conducted 20 minutes per day per group during a 5-day control period and during a 5-day experimental period when a concrete cylinder, approximately 1 m in length and 50 cm in diameter, was firmly placed in each enclosure. The mean incidence of agonistic interactions was 94 during the control condition versus only 45 during the experimental condition; the difference was statistically significant. The monkeys used the cylinders as escape routes to hide from potential aggressors.

Estep and Baker (1991) observed a breeding troop of 26 stump-tailed macaques during 90-minute sessions both before and after two solid temporary walls were erected within the animals' enclosure. The incidence of contact aggression was significantly lower when the monkeys had the option of breaking visual contact with other group members by moving behind these walls.

Ricker et al. (1995) hung sections of 12 cm diameter PVC pipes into the enclosure of heterogeneous groups of squirrel monkeys so that individuals could hide in them and break eye contact with potentially aggressive group members. This environmental modification *decreased fight wounds by 60 percent.*

Maninger et al. (1998) installed visual barriers in the living quarters of two breeding groups of 23 pig-tailed macaques and noted that the option of visual seclusion significantly reduced instances of biting, grabbing and chasing.

McCormack and Megna (2001) placed *privacy walls* in the enclosure of a 126-animal breeding troop of rhesus macaques and noted a significant decrease in threatening, chasing, fear grinning, and screaming.

3.3.2. Toys and Playrooms

Providing toys to captive primates might seem an obvious and simple solution to enriching their environment, though it is absolutely clear that no inanimate playthings can compare with the presence of a compatible conspecific. Whatever type of toy is provided, the attention span [of the animals] is limited to a day or two at most, and it is important to use any specific object [toy] only periodically, providing a constant variety to keep the animals interested (Dean, 1999). Nonhuman primates, just like human primates, are too intelligent not to quickly get bored by toys, unless these can gradually be destroyed. Not surprisingly, they are much more interested in destructible toys than in durable toys (chimpanzee: Shefferly et al., 1993; Brent and Stone, 1998; Videan et al., 2005b; orangutan: Heuer and Rothe, 1998; pig-tailed macaques: Cardinal and Kent, 1998).

Typically, new durable toys are removed after one or two days (cf., Anonymous, 2003) because they no longer elicit any interest but create a potential hygienic hazard (cf., Bayne et al., 1993). A conspicuous habituation to most commercial toys has been documented in chimpanzees (rubber and plastic toys for small children: Paquette and Prescott, 1988; Kong toys™: Pruetz and Bloomsmith, 1992; indestructible toy ball: Shefferly et al., 1993), rhesus macaques (nylon balls: Ross and Everitt, 1988; plastic toys for small children: Hamilton, 1991; nylon balls and rings, Kong toys™: Weick et al., 1991; Kong toys™: Bayne et al., 1993), baboons (nylon bones: Brent and Belik, 1997), long-tailed macaques (Kong toys™: Crockett et al., 1989) and pig-tailed macaques (plastic toys for small children: Cardinal and Kent, 1998; rubber and rawhide balls: Kessel and Brent, 1998). To be of some value for the animals, most commercial toys need to be replaced on a regular basis to make use of their short-lived novelty effect.

For baboons, Kong toys™ are an exception, and the animals show only moderate habituation to these flexible objects that seem to be particularly suitable for chewing. Group-housed female baboons who had continuous access to several Kong toys™ over a three-week period were *using* the toys consistently about 12 percent of 10-minute observations (Brent and Belik, 1997).

Kessel and Brent (1995) gave six female and six male young baboons, who lived alone in standard cages furnished with commercial toys (e.g., Kong toys™ and Nylarings™), access to a relatively large commercial toy-furnished play cage for two days each month. The animals' contact with toys increased significantly from about 4 percent of 10-minute observation sessions in the home cage to about 26 percent in the play cage.

Bryant et al. (1988) noted in six single-caged adult, male long-tailed macaques that the animals engaged in self-directed aggression, on average, 11 seconds per 30-minute observation. When each male had 30-minute access to a play cage daily, average duration of self-directed aggression was less than 3 seconds on the first and second day. No further self-directed aggression was observed thereafter until the end of the study at day 12.

3.3.3. Gnawing Sticks

Unlike many commercial toys, dry deciduous tree branches cut into *gnawing sticks* do not lose their novelty effect over time, since they steadily change their configuration and texture due to wear and progressive dehydration. The animals use the sticks for gnawing, nibbling, chewing, manipulating and playing. Long-term use of gnawing sticks by several hundred rhesus macaques resulted in no recognizable health hazards (Reinhardt, 1997a).

photo 128

Reinhardt (1990d) provisioned 20 adult pair-housed stump-tailed macaques each with a gnawing stick for two months. During a 60-minute observation session, 80 percent (16/20) of the animals gnawed the wooden material, on average, 8 percent of the time.

photo 129

Reinhardt (1990a) assessed the time budgets of 60 pair-housed rhesus macaques of both sexes. Each pair had continuous access to one regularly replaced gnawing stick for 18 months or longer. During two 30-minute remote video recordings, the gnawing stick was used by 94 percent (17/18) of the subadult animals versus 64 percent (27/42) of the adult animals. On average, subadults spent 10 percent, adults spent 3 percent of the time in direct contact with the stick.

Line and Morgan (1991) gave six adult female and six adult male, single-caged rhesus macaques each a gnawing stick and observed the animals during six 15-minute sessions spread over a period of four weeks. The sticks were used by the animals about 6 percent of the time.

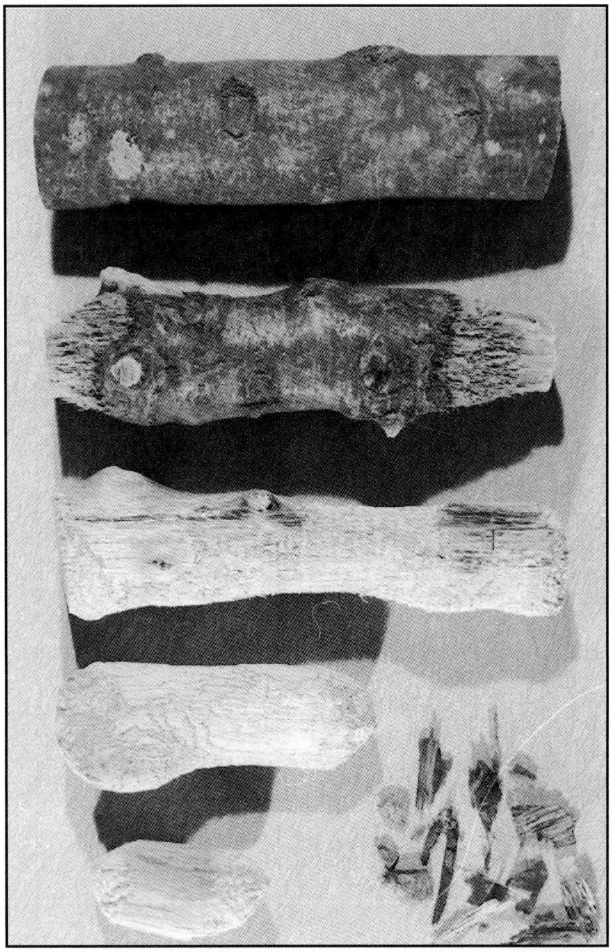

photo 130

Sticks of sun-dried red oak branches are particularly suitable because they gradually wear into flakes that are so small that even large quantities pass through the sewer drains without clogging (Reinhardt, 1992b).

3.3.4. Paper and Cardboard Boxes

Recycled paper and cardboard boxes are not expensive, but they can offer effective environmental enrichment for primates in small cages or larger enclosures.

Bryant et al. (1988) transferred six adult male long-tailed macaques from their standard home cages to a play pen, furnished with a telephone directory and a nylon ball, each day for 30 minutes over a 12-day test period. The animals had very little or no use for the nylon ball, but they spent, on average, 10 percent of the test sessions examining and shredding the telephone directory. Their interest in the paper material remained fairly constant; there was no indication that they lost interest in it over the course of time.

Beirise and Reinhardt (1992) placed a cardboard box into the pen of a 16-member breeding group of rhesus macaques once a week. After a habituation period of eight weeks, the animals were observed for two hours after placement of the cardboard box during week 9, 10 and 11. Individuals spent, on average, 65 percent of the two hours playing with the box, tearing it apart, shredding it and chewing pieces of it.

photo 131

Pruetz and Bloomsmith (1992) studied 22 chimpanzees of different age classes and both sexes who lived in pairs or small groups of up to five animals. They were all used to wrapper paper as part of the facility's enrichment program. In the course of a 16-week study, the paper was supplied one or two days per week. During the first 12 minutes right after paper distribution, the chimpanzees manipulated and played with the paper about 27 percent of the time without signs of habituation.

photo 132

Kessel et al. (1995) scattered shredded paper once a week throughout the room of a group of five young male chimpanzees. After a habituation period of one week, the animals were observed during 54-minute sessions, conducted Monday through Friday, during week two and three. They spent, on average, 27 percent of the time playing with the paper.

Smith et al. (2004) describe the case of an adolescent single-caged female chimpanzee who was over-grooming and picking at herself to the point of creating open lesions. The subject was offered large quantities of shredded paper on a continual basis. The hair pulling behavior decreased on the first day and continued to decrease in the course of a 12-week test period. It was concluded that the *provision of shredded paper has clearly shown to be a valuable tool when treating self-injurious behavior.*

3.3.5. Mirrors

Both apes and monkeys are fascinated by their own reflections and use a mirror to check out the immediate environment without directly looking at it (Gallup, 1970; Lethmate and Dücker, 1973; Eglash and Snowdon, 1983; Platt and Thompson, 1985; Anderson, 1986; Lambeth and Bloomsmith, 1992; O'Neill et al., 1997; Chiappa et al., 2004; de Waal et al., 2005; Schultz, 2006).

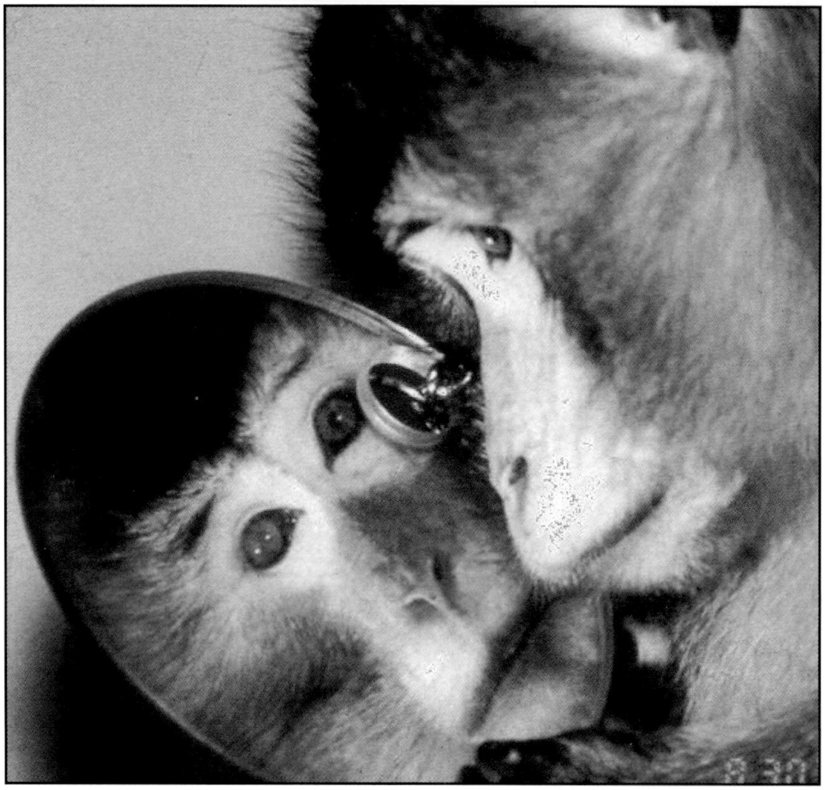

photo 133

Collinge (1989) exposed a heterogeneous group of six capuchin monkeys to a mirror attached from outside on the bar panel of the enclosure for a three-hour test period two times a week. Mirror-viewing time declined from an average of 24 percent per session in week one to 12 percent per session in week five. Four subadult animals spent considerably more time looking into the mirror than two adult animals.

Lambeth and Bloomsmith (1992) studied 20 adult and eight immature chimpanzees of both sexes who lived in 11 different enclosures either alone or in groups of four animals. Individuals were observed during two 12-minute sessions weekly, over a period of several weeks, when a 61-cm-diameter mirror was placed in front of the wire mesh wall of their enclosures. Housed in such a way that they had no visual contact with neighboring chimpanzees, they spent about 24 percent of the observation sessions in mirror-related behaviors, primarily staring at their own reflection. When the mirror gave visual access to neighboring chimpanzees, subjects spent about 30 percent of the time engaged in mirror-related expressions, gestures and behaviors. Immature chimpanzees interacted with the mirrors more than adults, who tended to gradually lose interest in them.

photo 134

Brent and Stone (1996) mounted mirrors on the outside of the enclosures of 13 female and seven male 6 to 17 years old chimpanzees who were housed alone or in pairs. Each subject was tested after having been exposed to the mirror for about two years. On average, individuals looked into the mirror—which could not be handled and adjusted to view other chimpanzees in the room—less than 1 percent of the time.

Mirrors that can be manipulated are particularly useful for animals who are housed alone, while socially housed animals tend to focus their attention more on the social partner than on the mirror. *Our singly housed baboons get the most enjoyment from their mirrors, while pair- and group-housed animals show little interest in them. We place the mirrors on the outside of the cages of our single-caged baboons, leave the mirrors only for a few hours at a time and replace them after a few days. This seems to work nicely: The animals' interest in the "new" mirror is always very strong, gradually declines and is hardly noticeable at the end of the day, when we take the "old" mirror away. Often the baboons will lip smack the mirrors or use them to look around the room. One boy was recently seen presenting to the mirror!* (LAREF, 2007e).

Harris and Edwards (2004) hung stainless steel, 15-cm diameter mirrors on the cages' front panels of 25 single male vervet monkeys and observed each subject during four 30-minute sessions, 10 months, and again 16 months after the initial introduction of the mirrors. The average time spent contacting the mirror and looking into the mirror was consistent at about five percent, indicating that the animals had a sustained interest in them.

photo 135

We have found an acrylic sheet mirror that we can cut into different-size pieces. Some get hung on the walls, using double-sided tape, while other pieces get hung right inside the enclosures, using zip ties. We also cut small pieces and give these directly to the primates. Our rhesus macaques often combine the wall and hand mirrors to get extra viewing advantage! It is really fun to watch them. The acrylic leaves no sharp edges when it breaks; this means it is safe for the animals. We never encountered a problem (LAREF, 2007e).

photo 136

3.3.6. Windows and Light

Indoor-housed primates are often locked up in quarters with no exterior windows. The situation is particularly grave for animals kept in the dark environment of lower-tier cages.

Exterior windows can ameliorate this situation. The International Guidelines for the Acquisition, Care and Breeding of Nonhuman Primates recommend that:

> *Whenever possible, rooms housing nonhuman primates should be provided with windows, since they are a source of natural light and can provide health benefits as well as environmental enrichment* (International Primatological Society, 2007). *Windows through which the animals can see the outside world may help to alleviate some boredom* (Primate Research Institute, 2003).

We expose our squirrel monkeys to natural daylight via big windows during the summer. This is supplemented with artificial light in late fall and early spring, when the days are short, and throughout the winter. Some of our squirrel monkeys will lie as close to the window as possible and let the sun rays dance on their bell (LAREF, 2007f).

photo 137

I've seen the same behavior in our marmosets. As soon as the sunlight hits the window, the animals stop what they are doing, run over to the window ledge, and start stretching out and basking in the sunrays. There is no doubt in my mind that exposure to natural light, especially sunlight, is highly appreciated by the animals.

All our rhesus macaques have access to one-way glass exterior windows mounted high above ground level. I very often see the animals gather up, attentively gazing out of the windows towards the source of some noise, at caretakers, activities in the garden and birds (LAREF, 2007f).

photo 138

Pairs of male long-tailed macaques, transferred regularly for 90-minute periods to a playroom with windows, spent about 67 percent of their time looking out the windows (Lynch and Baker, 2000).

In a study of chimpanzees, it was found that regardless of enclosure size or group composition, all subjects preferred locations close to windows with visual access to the outdoors and/or caregiver maintenance activity (Fritz et al., 1992).

There seems to be an international regulatory and professional consensus that:

> *Lighting* ***must*** [emphasis added] *be uniformly diffused throughout animal facilities and provide sufficient illumination to aid in maintaining good housekeeping practices, adequate cleaning, adequate inspection for animals, and for the well-being of the animals* (United States Department of Agriculture, 1991; cf., Institute of Laboratory Animal Resources, 1980; National Research Council, 1996; Fortman et al., 2002; International Primatological Society, 2007).

These important stipulations are meaningless as long as the traditional double-tier caging system prevails in some countries, such as the United States (Rosenberg and Kesel, 1994). *The sanitation tray, which runs the length of the room beneath the upper tier of cages, reduces significantly the amount of light from ceiling-mounted fixtures that can penetrate to the lower cage tier; animals in the lower tier are thus relegated to a permanent state of semi-gloom* (Mahoney, 1992). Illumination is often so poor that flashlights are needed to identify animals, check their well-being, and make sure that the bottom of the cage is adequately cleaned (Reinhardt, 1997b; Reasinger and Rogers, 2001).

The original cages used for housing monkeys individually [in the USA] *were modified chicken or turkey cages* (Kelley and Hall, 1995) stacked on top of each other in rows of two or three tiers. This caging system was introduced in the 1950s to quickly provide short-term accommodation for hundreds of thousands of monkeys used for the development of vaccines. Today, there is no longer a need for such large numbers of animals, but the much smaller numbers still used remain stuck in this outdated caging system not because there is an emergency, but because it saves money to house twice or thrice the number of animals in multi-tier racks instead of in single rows.

photo 139

It is surprising that cage location of research animals is rarely mentioned in scientific articles (Reinhardt and Reinhardt, 2000), although the environment of upper- and lower-tier housed animals markedly differs in terms of illumination and living dimension, i.e., terrestrial in bottom row versus arboreal in top row. Not accounting for these important variables is likely to increase data variability and, consequently, the number of experimental animals needed to obtain statistically significant results (Home Office, 1989; Institute for Laboratory Animal Research, 1992; Russell and Burch, 1959).

Rotating cage positions relative to the light source—as is sometimes recommended (Canadian Council on Animal Care, 1993; National Research Council, 1996) and often practiced (Ott, 1974; Ross and Everitt, 1988; Shively, 2001; Buchanan-Smith et al., 2002)—rotates the inherent problem, but it does not solve it: There will always be half of a population of double-tier caged animals who live in the lower tier in the shade cast by the cages of the upper tier.

photo 140

Keeping nonhuman primates in single-tier, rather than multi-tier, caging systems in high cages equipped with high perches, verandas and shelves is, at the moment, the only satisfactory refinement option. It:

- provides all animals of the room uniform illumination,
- creates uniform illumination to aid in maintaining good housekeeping practices, adequate cleaning and adequate inspection for animals, and
- allows the animals to access the "arboreal" dimension of their enclosures and retreat to "safe" vantage points above eye-level of attending personnel.

3.3.7. Videos and Television

Bloomsmith et al. (1990) recorded four times the responses of three single-caged adult female and one single-caged subadult male chimpanzees to the blank monitor screen and to videotapes of other animals and humans; each test lasted 20 minutes. Subjects watched the monitor, on average, 7 percent of the time when no videotape played, versus 74 percent of the time when a videotape was shown. The content of the videos did not affect the animals' interest in them.

Brent and Stone (1996) exposed 13 adult female and seven subadult male chimpanzees daily for approximately six hours to commercial television programs. The animals lived alone or in pairs when they were tested after about two years of watching television. Individual chimpanzees looked at the television, on average, about 2 percent of the time; housing condition did not influence watching time. The authors did not test the animals when the TV screen was blank to see if the content actually mattered to them.

photo 141

Schapiro and Bloomsmith (1995) presented 49 single-caged yearling rhesus macaques with videotapes of chimpanzees and rhesus macaques in natural settings most of the day for a period of three months. During several 15-minute observation sessions, subjects were looking at the monitor about 3 percent of the time. The possibility was not ruled out that the animals would have shown the same interest in the blank monitor.

3.3.8. Music

Brent and Weaver (1996) studied four subadult single-caged baboons who were used to having a radio station playing "oldies" throughout the day. They showed a significant increase in mean heart rate whenever the radio was turned off for short periods of time, probably because they were so used to it that any disruption of the music was an uncomfortable experience for them.

Hinds et al. (2007) exposed nine single-caged vervet monkeys who were not used to music to a 90-minute period of recorded harp music and noted no change in heart rate, blood pressure, respiratory rate and body temperature.

Videan et al. (2007) broadcasted various types of music over an intercom system that could be heard by 31 female and 26 male chimpanzees living in groups of 3 to 7 animals. The animals were observed for one hour before music was turned on and again for one hour while music was playing. Listening to instrumental music significantly increased affiliative behavior in both sexes, while listening to vocal music significantly decreased agonistic behavior in males, but not in females.

Markowitz and Line (1989) and Line et al. (1990b) mounted a radio device on the cages of five single adult female rhesus macaques. The radio was available for a 7-week period and was preset to a "soft rock" format station; the animals could turn the radio on and off by touching two different bars. On average, they had it turned on approximately 50 percent the 24-hour day, and they showed no signs of losing interest in listening to the music.

McDermott and Hauser (2007) gave four adult cotton-top tamarins and four adult common marmosets the choice of listening to various noises and various kind of music. The animals showed a significant preference for soft over loud noise and for slow tempo over fast tempo music. *Both tamarins and marmosets strongly and consistently preferred silence over musical stimuli* (flute lullaby: $p<0.0001$; sung lullaby: $p<0.003$; Mozart concerto: $p<0.0001$), *suggesting that they did not find such stimuli pleasurable or relaxing.*

3.3.9. Water

Basins filled with water for swimming, diving for food items, fishing for food items, and playing have been employed for caged and group-housed long-tailed macaques (Gilbert and Wrenshall, 1989), squirrel monkeys (King and Norwood, 1989) and rhesus macaques (Anderson et al., 1992; Rawlins, 2005) without adverse effects, other than much splashing.

We give our pair-housed cynos "bathtubs," filled with 30 to 40 cm deep warm water, a few times a week, and have never encountered any problems other than a lot of splashing. Some monkeys take luxurious baths, others climb a perch and jump into the water, others sit on the side walls and drag their hands in the water, and others wash their fruit in the water. Usually the monkeys make a real mess within the first half hour, and yes they do urinate/defecate in the water. We empty the tubs after about two hours, if the monkeys haven't done it already themselves—which is often the case (LAREF, 2007g).

photo 142

4. CONCLUSIONS

There is no justification—other than veterinary reasons and social incompatibility—to keep nonhuman primates in barren cages and to distress them during common handling procedures. Species-adequate, effective and practicable options for providing environmental enrichment and practicable options of training nonhuman primates to cooperate during common procedures have been described, tested and documented in the scientific and professional literature. Making life easier for nonhuman primates in research laboratories is not only an ethical and animal welfare priority, but a fundamental condition for the scientific validity of the research data collected from these animals (Animal Welfare Institute 1979; National Research Council, 1985; Meyerson, 1986; Donnelley, 1990; Morton, 1990; Novak and Bayne, 1991; Schwindaman, 1991; Institute for Laboratory Animal Research, 1992; Chance and Russell, 1997; Fuchs, 1997; Öbrink and Rehbinder, 1999; Richmond, 2002; Reinhardt and Reinhardt, 2002; Russell, 2002).

photo 143

A good management program provides the environment, housing, and care that minimizes variations that can affect research. Animals should be housed with the goal of maximizing species-specific behaviors and minimizing stress-induced behaviors (National Research Council, 1996).

The maintenance and use of non-human primates should only be permitted in facilities which can truly provide the high quality of housing, and care and attention which these animals require, if their normal physiology and behaviour are to be maintained (Balls, 1995).

photo 144

Sharing the same roots makes it easy for any compassionate *human* primate to alleviate the suffering of a *nonhuman* primate who is imprisoned, subjected to life-threatening research procedures and, finally, sentenced to death.

This applies not only to animal technicians and animal caretakers but particularly to veterinarians who *pledge to take responsibility for the welfare of animals* [and] *vow to use scientific knowledge and skills for the advancement of medical knowledge. The wise composer of this oath saw no conflict between relieving animal suffering and advancing science. Indeed there is none* (Schwindaman, 1991).

5. REFERENCES

Abee CR 1985 Medical care and management of the squirrel monkey. In: Rosenblum LA and Coe CL (ed) *Handbook of Squirrel Monkey Research* pp. 447-488. Plenum Press: New York, NY

Abney DM and Weed JL 2006 Methods for successfully pair housing adult male rhesus macaques (*Macaca mulatta*). *American Journal of Primatology 68*(Supplement): 59

Ackerley ET and Stones PB 1969 Safety procedures for handling monkeys. *Laboratory Animal Handbooks 4*: 207-211

Alexander S and Fontenot MB 2003 Isosexual social group formation for environmental enrichment in adult male *Macaca mulatta. AALAS [American Association for Laboratory Animal Science] 54th National Meeting Official Program*: 141

Alford PL, Bloomsmith MA, Keeling ME and Beck TF 1995 Wounding aggression during the formation and maintenance of captive, multimale chimpanzee groups. *Zoo Biology 14*: 347-359

Altman NH 1970 Restraint of monkeys in clinical examination and treatment. *Journal of the American Veterinary Medical Association 159*: 1222

Animal Welfare Institute 1979 *Comfortable Quarters for Laboratory Animals, Seventh Edition*. Animal Welfare Institute: Washington, DC

American Association for Laboratory Animal Science 2001 *Cost of Caring: Recognizing Human Emotions in the Care of Laboratory Animals*. American Association for Laboratory Animal Science: Memphis, TN
http://www.aalas.org/pdf/06-00006.pdf

Anderson JR 1986 Mirror-mediated finding of hidden food by monkeys *(Macaca tonkeana* and *Macaca fascicularis). Journal of Comparative Psychology 100*: 237-242

Anderson JR and Chamove AS 1984 Allowing captive primates to forage. *Standards in Laboratory Animal Management. Proceedings of a Symposium* pp. 253-256. The Universities Federation for Animal Welfare: Potters Bar, UK

Anderson JR, Peignot P and Adelbrecht C 1992 Task-directed and recreational underwater swimming in captive rhesus monkeys *(Macaca mulatta). Laboratory Primate Newsletter 31*(4): 1-4
http://www.brown.edu/Research/Primate/lpn31-4.html#swim

Anonymous 2003 Enrichment survey results. *Animal Keeper's Forum 30*: 513-515

Anzenberger G and Gossweiler H 1993 How to obtain individual urine samples from undisturbed marmoset families. *American Journal of Primatology 31*: 223-230

Asvestas C 1998 Pairing *Macaca fascicularis. Laboratory Primate Newsletter 37*(3): 5
http://www.brown.edu/Research/Primate/lpn37-3.html#wao

Asvestas C and Reininger M 1999 Forming a bachelor group of long-tailed macaques *(Macaca fascicularis). Laboratory Primate Newsletter 38*(3): 14
http://www.brown.edu/Research/Primate/lpn38-3.html#group

Aureli F, Veenema H and van Eck C 1993 Coping with short-term crowding in long-tailed macaques. *American Journal of Primatology. 31*: 295

Aureli F, Preston SD and de Waal FBM 1999 Heart rate responses to social interactions in free-moving rhesus macaques: (Macaca mulatta): A pilot study. *Journal of Comparative Psychology 113*(1): 59-65

Baker KC 1997 Human interaction as enrichment for captive chimpanzees: A preliminary report. *American Journal of Primatology 42*: 92

Baker KC 2004 Benefits of positive human interaction for socially housed chimpanzees. *Animal Welfare 13*: 239-245

Balls M 1995 The use of non-human primates as laboratory animals in Europe: Moving toward the zero option. *ATLA [Alternatives to Laboratory Animals] 23*: 284-286

Basile BM, Hampton RR, Chaudhry AM and Murray EA 2007 Presence of a privacy divider increases proximity in pair-housed rhesus monkeys. *Animal Welfare 16*(1): 37-39

Bayne K 2002 Development of the human-research animal bond and its impact on animal well-being. *ILAR [Institute for Laboratory Animal Research] Journal 43*(1): 4-9
http://dels.nas.edu/ilar_n/ilarjournal/43_1/Development.shtml

Bayne K, Dexter SL, Mainzer H, McCully C, Campbell G and Yamada F 1992 The use of artificial turf as a foraging substrate for individually housed rhesus monkeys *(Macaca mulatta). Animal Welfare 1*: 39-53
http://www.awionline.org/Lab_animals/biblio/aw1-39.htm

Bayne K, Hurst JK and Dexter SL 1992 Evaluation of the preference to and behavioral effects of an enriched environment on male rhesus monkeys. *Laboratory Animal Science 42*: 38-45

Bayne K, Dexter SL, Hurst JK, Strange GM and Hill EE 1993 Kong toys for laboratory primates: Are they really an enrichment or just fomites? *Laboratory Animal Science 43*: 78-85

Bayrakci R 2003 Starting an injection training program with lion-tailed macaques (Macaca silenus). *Animal Keeper's Forum 30*: 503-512

Beckley SA and Novak MA 1989 Examination of various foraging components and their suitability as enrichment tools for captively housed primates. *American Journal of Primatology 19*(Supplement): 37-43

Beirise JH and Reinhardt V 1992 Three inexpensive environmental enrichment options for group-housed *Macaca mulatta. Laboratory Primate Newsletter 31*(1): 7-8
http://www.brown.edu/Research/Primate/lpn31-1.html#three

Bell BK 1995 Dealing with specific behavioral problems using operant conditioning with bonobos *(Pan paniscus). American Zoo and Aquarium Association (AZA) Regional Conference Proceedings*: 26-28

Bennett BT 1990 Alternative methodologies. In: Bennett BT, Brown MJ and Schofield JC (eds) *Essentials for Animal Research: A Primer for Research Personnel* pp. 13-25. United States Department of Agriculture, National Agricultural Library: Beltsville, MD

Bentson KL, Capitanio JP and Mendoza SP 2003 Cortisol responses to immobilization with Telazol or ketamine in baboons (*Papio cynocephalus/anubis*) and rhesus macaques (*Macaca mulatta*). *Journal of Medical Primatology 32*: 148-160

Bernstein IS and Mason WA 1963 Group formation by rhesus monkeys. *Animal Behaviour 11*: 28-31

Bertrand F, Seguin Y, Chauvier F and Blanquié JP 1999 Influence of two different kinds of foraging devices on feeding behaviour of rhesus macaques *(Macaca mulatta). Folia Primatologica 70*: 207

Bjone SJ, Price IR and McGreevy PD 2006 Food distribution effects on the behaviour of captive common marmosets, *Callithrix jacchus. Animal Welfare 15*: 131-140

Blois-Heulin C and Jubin R 2004 Influence of the presence of seeds and litter on the behaviour of captive red-capped mangabeys *Cercocebus torquatus torquatus. Applied Animal Behaviour Science 85*: 340-362

Bloom KR and Cook M 1989 Environmental enrichment: Behavioral responses of rhesus to puzzle feeders. *Lab Animal 18*(5): 25,27,29,31

Bloomsmith MA, Alford PL and Maple TL 1988 Successful feeding enrichment for captive chimpanzees. *American Journal of Primatology 16*: 155-164

Bloomsmith MA, Keeling ME and Lambeth SP 1990 Videotapes: Environmental enrichment for singly housed chimpanzees. *Lab Animal 19*(1): 42-46

Bloomsmith MA, Stone AM and Laule GE 1998 Positive reinforcement training to enhance the voluntary movement of group-housed chimpanzees within their enclosure. *Zoo Biology 17*: 333-341

Boccia ML 1989a Comparison of the physical characteristics of grooming in two species of macaques (*Macaca nemestrina* and *M. radiata*). *Journal of Comparative Psychology 103*(2): 177-183

Boccia ML 1989b Preliminary report on the use of a natural foraging task to reduce aggression and stereotypies in socially housed pigtail macaques. *Laboratory Primate Newsletter 28*(1): 3-4
http://www.brown.edu/Research/Primate/lpn28-1.html#maria

Boccia ML and Hijazi AS 1998 A foraging task reduces agonistic and stereotypic behaviors in pigtail macaque social groups. *Laboratory Primate Newsletter 37*(3): 1-5
http://www.brown.edu/Research/Primate/lpn37-3.html#boccia

Boccia ML, Reite M and Laudenslager ML 1989 On the physiology of grooming in a pigtail macaque. *Physiology and Behavior 45*: 667-670

Boinski S, Noon C, Stans S, Samudio R, Sammarco P and Hayes A 1994 The behavioral profile and environmental enrichment of a squirrel monkey colony. *Laboratory Primate Newsletter 33*(4): 1-4
http://www.brown.edu/Research/Primate/lpn33-4.html#boin

Bourgeois SR and Brent L 2005 Modifying the behaviour of singly caged baboons: evaluating the effectiveness of four enrichment techniques. *Animal Welfare 14*: 71-81

Brent L and Eichberg JW 1991 Primate puzzleboard: A simple environmental enrichment device for captive chimpanzees. *Zoo Biology 10*: 353-360

Brent L and Long KE 1995 The behavioral response of individually caged baboons to feeding enrichment and the standard diet: A preliminary report. *Contemporary Topics in Laboratory Animal Science 34*(2): 65-69

Brent L and Stone AM 1996 Long-term use of television, balls, and mirrors as enrichment for paired and singly caged chimpanzees. *American Journal of Primatology 39*: 139-145

Brent L and Weaver D 1996 The physiological and behavioral effects of radio music on singly housed baboons. *Journal of Medical Primatology 25*: 370-374
http://www.awionline.org/Lab_animals/biblio/jmp25-3.htm

Brent L and Belik M 1997 The response of group-housed baboons to three enrichment toys. *Laboratory Animals 31* : 81-85

Brent L and Stone AM 1998 Destructible toys as enrichment for captive chimpanzees. *Journal of Applied Animal Welfare Science 1*: 5-14
http://www.psyeta.org/jaaws/abv1n1.html

Brown CS and Loskutoff NM 1998 A training program for noninvasive semen collection in captive western lowland gorillas *(Gorilla gorilla gorilla). Zoo Biology 17*: 143-151

Bryant CE, Rupniak NMJ and Iversen SD 1988 Effects of different environmental enrichment devices on cage stereotypies and autoaggression in captive cynomolgus monkeys. *Journal of Medical Primatology 17*: 257-269
http://www.awionline.org/Lab_animals/biblio/jmp17-2.htm

Buchanan-Smith HM 1991 A field study on the red-bellied tamarin, *Saguinus l. labiatus,* in Boliva. *International Journal of Primatology 12*: 259-276

Buchanan-Smith HM, Shand C and Morris K 2002 Cage use and feeding height preferences of captive common marmosets (*Callithrix j. jacchus*) in two-tier cages. *Journal of Applied Animal Welfare Science 5*: 139-149

Bunyak SC, Harvey NC, Rhine RJ and Wilson MI 1982 Venipuncture and vaginal swabbing in an enclosure occupied by a mixed-sex group of stumptailed macacaques *(Macaca arctoides). American Journal of Primatology 2*: 201-204

Byrum R and St. Claire M 1998 Pairing female *Macaca nemestrina. Laboratory Primate Newsletter 37*(4): 1
http://www.brown.edu/Research/Primate/lpn37-4.html#byrum

Canadian Council on Animal Care 1993 *Guide to the Care and Use of Experimental Animals, Volume 1, 2nd Edition.* Canadian Council on Animal Care: Ottawa, Canada
http://www.ccac.ca/en/CCAC_Programs/Guidelines_Policies/GUIDES/ENGLISH/toc_v1.htm

Canadian Council on Animal Care 1984 Chapter XX: Non-human primates. In: *Guide to the Care and Use of Experimental Animals, Volume 2* pp. 163-173. Canadian Council on Animal Care: Ottawa, Canada

Cardinal BR and Kent SJ 1998 Behavioral effects of simple manipulable environmental enrichment on pair-housed juvenile macaques (*Macaca nemestrina*). *Laboratory Primate Newsletter 37*(1): 1-3
http://www.brown.edu/Research/Primate/lpn37-1.html#kent

Caws C and Aureli F 2001 Coping with short-term space restriction in chimpanzees. *Primate Eye [Primate Society of Great Britain Spring Meeting] 74*: 9
http://www.psgb.org/Meetings/Spring2001.html

Celli ML, Tomonagaa M, Udonob T, Teramotob M and Naganob K 2003 Tool use task as environmental enrichment for captive chimpanzees. *Applied Animal Behaviour Science 81*: 171-182

Chamove AS and Scott L 2005 Forage box as enrichment in single- and group-housed callitrichid monkeys. *Laboratory Primate Newsletter 44*(2): 13-17
http://www.brown.edu/Research/Primate/lpn44-2.html#box

Chance MRA and Russell WMS 1997 The benefits of giving experimental animals the best possible environment. In: Reinhardt V (ed) *Comfortable Quarters for Laboratory Animals, Eighth Edition* pp. 12-14. Animal Welfare Institute: Washington, DC

Chiappa P, Ortiz-Sánchez V and Antonio-Garcés J 2004 Social status and mirror behaviour in *Macaca arctoides. Folia Primatologica 75*(Supplement): 364

Chopra PK, Seth PK and Seth S 1992 Behavioural profile of free-ranging rhesus monkeys. *Primate Report 32*: 75-105

Clarence WM, Scott JP, Dorris MC and Paré M 2006 Use of Enclosures with Functional Vertical Space by Captive Rhesus Monkeys (*Macaca mulatta*) Involved in Biomedical Research. *JAALAS [Contemporary Topics in Laboratory Animal Science] 45*(5): 31-34

Clarke AS, Juno CJ and Maple TL 1982 Behavioral effects of a change in the physical environment: a pilot study of captive chimpanzees. *Zoo Biology 1*: 371-380

Clarke AS, Czekala NM and Lindburg DG 1995 Behavioral and adrenocortical responses of male cynomolgus and lion-tailed macaques to social stimulation and group formation. *Primates 36*: 41-46

Coe CL and Rosenblum LA 1984 Male dominance in the bonnet macaque: A malleable relationship. In: Barchas PR and Mendoza SP (eds) *Social Cohesion. Essays Toward a Sociophysiological Perspective* pp. 31-64. Greenwood Press: Westport, CT

Coe CL, Franklin D, Smith ER and Levine S 1982 Hormonal responses accompanying fear and agitation in the squirrel monkey. *Physiology and Behavior 29*: 1051-1057

Coelho AM and Carey KD 1990 A social tethering system for non-human primates used in laboratory research. *Laboratory Animal Science 40*: 388-394

Coelho AM, Carey KD and Shade RE 1991 Assessing the effects of social environment on blood pressure and heart rates of baboons. *American Journal of Primatology 23*: 257-267

Collinge NE 1989 Mirror reactions in a zoo colony of cebus monkeys. *Zoo Biology 8*: 89-98

Council of Europe 2006 *Appendix A of the European Convention for the Protection of Vertebrate Animals Used for Experimental and Other Scientific Purposes (ETS No. 123)*. Council of Europe: Strasbourg, France
http://conventions.coe.int/Treaty/EN/Treaties/PDF/123-Arev.pdf

Crockett CM and Heffernan KS 1998 Grooming-contact cages promote affiliative social interaction in individually housed adult baboons. *American Journal of Primatology 45*: 176

Crockett CM, Bielitzki JT, Carey A and Velez A 1989 Kong toys as enrichment devices for singly-caged macaques. *Laboratory Primate Newsletter 28*(2): 21-22
http://www.brown.edu/Research/Primate/lpn28-2.html#kong

Crockett CM, Bowers CL, Bowden DM and Sackett GP 1994 Sex differences in compatibility of pair-housed adult longtailed macaques. *American Journal of Primatology 32*: 73-94

Crockett CM, Bellanca RU, Bowers CL and Bowden DM 1997 Grooming-contact bars provide social contact for individually caged laboratory primates. *Contemporary Topics in Laboratory Animal Science 36*(6): 53-60

Crockett CM, Koberstein D and Heffernan KS 2001 Compatibility of laboratory monkeys housed in grooming-contact cages varies by species and sex. *American Journal of Primatology 54*(Supplement): 51-52
http://www.asp.org/asp2001/abstractDisplay.cfm?abstractID=112&confEventID=26

Crockett CM, Lee GH and Thom JP 2006 Sex and age predictors of compatibility in grooming-contact caging vary by species of laboratory monkey. *International Journal of Primatology 27*(Supplement): 417

Cross N, Pines MK and Rogers LJ 2004 Saliva sampling to assess cortisol levels in unrestrained common marmosets and the effect of behavioral stress. *American Journal of Primatology 62*: 107-114

de Waal FBAF 1997 Conflict resolution and distress alleviation in monkeys and apes. *Annals of the New York Academy of Science 807*: 317-328

de Waal FBM and Ren RM 1988 Comparison of the reconciliation behavior of stumptail and rhesus macaques. *Ethology 78*: 129-142

de Waal FBM, Dindo M, Freeman CA and Hall MJ 2005 The monkey in the mirror: Hardly a stranger. *Proceedings of the National Academy of Sciences 102*: 140-147
http://www.pnas.org/cgi/content/full/102/32/11140#REF3

Dean SW 1999 Environmental enrichment of laboratory animals used in regulatory toxicology studies. *Laboratory Animals 33*: 309-327

Dettling A 1997 Physical environment and its influence on behaviour in captive common marmosets *(Callithrix jacchus)*. In: Pryce C, Scott L and Schnell C (eds) *Marmosets and Tamarins in Biological and Biomedical Research. Proceedings of a Workshop* pp. 54-59. DSSD Imagery: Salisbury, UK

Dexter SL and Bayne K 1994 Results of providing swings to individually housed rhesus monkeys (*Macaca mulatta*). *Laboratory Primate Newsletter 33*(2): 9-12
http://www.brown.edu/Research/Primate/lpn33-2.html#bayne

Donnelly MJ, Wickham A, Kulick A, Rogers I, Stribling S, Strack A, Doerning B and Feeney W 2007 A refinement of oral dosing in the common marmoset (*Callithrix jacchus*). *AALAS [American Association for Laboratory Animal Science] 58th National Meeting Official Program*: 45

Donnelley S 1990 Animals in science: The justification issue. In: Donnelley S and Nolan K (eds) *Animal, Science and Ethics* pp. 8-13. The Hastings Center Report: Briarcliff Manor, CT

Doyle LA, Baker KC and Cox LD 2008 Physiological and behavioral effects of social introduction on adult male rhesus macaques. *American Journal of Primatology 70*: 1-9

Eaton GG, Kelley ST, Axthelm MK, Iliff-Sizemore SA and Shiigi SM 1994 Psychological well-being in paired adult female rhesus (*Macaca mulatta*). *American Journal of Primatology 33*: 89-99

Eglash AR and Snowdon CT 1983 Mirror-image responses in pygmy marmosets *(Cebuella pygmaea)*. *American Journal of Primatology 5*: 211-219

Elvidge H, Challis JRG, Robinson JS, Roper C and Thorburn GD 1976 Influence of handling and sedation on plasma cortisol in rhesus monkeys *(Macaca mulatta)*. *Journal of Endocrinology 70*: 325-326

Erwin J 1977 Factors influencing aggressive behavior and risk of trauma in the pigtail macaque *(Macaca nemestrina). Laboratory Animal Science 27*: 541-547

Erwin J, Anderson B, Erwin N, Lewis L and Flynn D 1976 Aggression in captive pigtail monkey groups: Effects of provision of cover. *Perceptual and Motor Skills 42*: 319-324

Estep DQ and Baker SC 1991 The effects of temporary cover on the behavior of socially housed stumptailed macaques (*Macaca arctoides*). *Zoo Biology 10*: 465-472

Fekete JM, Norcross JL and Newman JD 2000 Artificial turf foraging boards as environmental enrichment for pair-housed female squirrel monkeys. *Contemporary Topics in Laboratory Animal Science 39*(2): 22-26

Fortman, JD, Hewett, TA and Bennett, BT 2002 *The Laboratory Nonhuman Primate.* CRC Press: Boca Raton, FL

Fragaszy DM, Baer J and Adams-Curtis LE 1994 Introduction and integration of strangers into captive groups of tufted capuchins *(Cebus apella). International Journal of Primatology 15*: 399-420

Fritz J 1989 Resocialization of captive chimpanzees: An amelioration procedure. *American Journal of Primatology 19*(Supplement): 79-86

Fritz J 1994 Introducing unfamiliar chimpanzees to a group or partner. *Laboratory Primate Newsletter 33*(1): 5-7 http://www.brown.edu/Research/Primate/lpn33-1.html#jo

Fritz P and Fritz J 1979 Resocialization of chimpanzees: Ten years of experience at the Primate Foundation of Arizona. *Journal of Medical Primatology 8*: 202-221

Fuchs E 1997 Requirements of biomedical research in terms of housing and husbandry: Neuroscience. *Primate Report 49*: 43-46

Gallup GG 1970 Chimpanzees: Self-recognition. *Science 167*: 86-87

Gantt WH, Newton JEO, Royer FL and Stephens JH 1966 Effect of person. *Conditional Reflex 1*: 18-35

Gilbert SG and Wrenshall E 1989 Environmental enrichment for monkeys used in behavioral toxicology studies. In: Segal EF (ed) *Housing, Care and Psychological Wellbeing of Captive and Laboratory Primates* pp. 244-254. Noyes Publications: Park Ridge, NJ

Gilloux I, Gurnell J and Shepherdson D 1992 An enrichment device for great apes. *Animal Welfare 1*: 279-289 http://www.awionline.org/Lab_animals/biblio/aw1-279.htm

Gisler DB, Benson RE and Young RJ 1960 Colony husbandry of research monkeys. *Annals of the New York Academy of Sciences 85*: 758-568

Goff C, Howell SM, Fritz J and Nankivell B 1994 Space use and proximity of captive chimpanzees (*Pan troglodytes*) mother/offspring pairs. *Zoo Biology 13*: 61-68

Gonzalez CA, Coe CL and Levine S 1982 Cortisol responses under different housing conditions in female squirrel monkeys. *Psychoneuroendocrinology 7*: 209-216

Goodwin J 1997 The application, use, and effects of training and enrichment variables with Japanese snow macaques (*Macaca fuscata*) at the Central Park Wildlife Center. *American Zoo and Aquarium Association (AZA) Regional Conference Proceedings*: 510-515

Grief L, Fritz J and Maki S 1992 Alternative forage types for captive chimpanzees. *Laboratory Primate Newsletter 31*(2): 11-13 http://www.brown.edu/Research/Primate/lpn31-2.html#grief

Günther MM 1998 Influence of habitat structure on jumping behaviour in *Galago moholi*. *Folia Primatologica 69*(Supplement): 410

Gunnar MR, Gonzalez CA and Levine S 1980 The role of peers in modifying behavioral distress and pituitary-adrenal response to a novel environment in year-old rhesus monkeys. *Physiology and Behavior 25*: 795-798

Gust DA, Gordon TP, Wilson ME, Ahmed-Ansari A, Brodie AR and McClure HM 1991 Formation of a new social group of unfamiliar female rhesus monkeys affects the immune and pituitary adrenocortical systems. *Brain, Behavior, and Immunity 5*: 296-307

Gust DA, Gordon TP, Hambright MK and Wilson ME 1993 Relationship between social factors and pituitary-adroneocortical activity in female rhesus monkeys (*Macaca mulatta*). *Hormones and Behavior 27*: 318-331

Gust DA, Gordon TP, Brodie AR and McClure HM 1994 Effect of a preferred companion in modulating stress in adult female rhesus monkeys. *Physiology and Behavior 55*: 681-684

Gust DA, Gordon TP, Wilson ME, Brodie AR, Ahmed-Ansari A and McClure HM 1996 Group formation of female pigtail macaques (*Macaca nemestrina*). *American Journal of Primatology 39*: 263-273

Hall KRL and De Vore I 1965 Baboon social behavior. In: De Vore I (ed) *Primate Behavior - Field Studies of Monkeys and Apes* pp. 53-110. Holt, Rinehart and Winston: New York, NY

Hamilton P 1991 Enrichment toys and tools in recent trials. *Humane Innovations and Alternatives in Animal Experimentation 5*: 272-277

Hanya G 2004 Seasonal variations in the activity budget of Japanese macaques the coniferous forest of Yakushima: Effects of food and temperature. *American Journal of Primatology 63*: 164-177

Harris HG and Edwards AJ 2004 Mirrors as environmental enrichment for African green monkeys . *American Journal of Primatology 63*: 459-467

Hayes SL 1990 Increasing foraging opportunities for a group of captive capuchin monkeys (*Cebus capucinus*). *Laboratory Animal Science 40*: 515-519

Hennessy MB 1984 Presence of companion moderates arousal of monkeys with restricted social experience. *Physiology and Behavior 33*: 393-398

Henrickson RV 1976 The nonhuman primate. *Lab Animal 5*(4): 60-62

Heuer A and Rothe H 1998 Environmental enrichment for four subadult orangutans at the Hannover Zoo [article in German]. *Der Zoologische Garten 2*: 119-133

Hinds SB, Raimond S and Purcell BK 2007 The effect of harp music on heart rate, mean blood pressure, respiratory rate, and body temperature in the African green monkey. *Journal of Medical Primatology 36*(2): 95-100

Home Office 1989 *Animals (Scientific Procedures) Act 1986. Code of Practice for the Housing and Care of Animals Used in Scientific Procedures*. Her Majesty's Stationery Office: London, UK
http://scienceandresearch.homeoffice.gov.uk/animal-research/legislation/

Howell SM, Miteva E, Fritz J and Baron J 1997 The provision of cage furnishings as environmental enrichment at the Primate Foundation of Arizona. *The Newsletter 9*(2): 1-5
http://www.awionline.org/Lab_animals/biblio/jo-5.htm

Hrapkiewicz, K, Medina, L and Holmes, DD 1998 *Clinical Medicine of Small Mammals and Primates, Second Edition.* Manson Publishing: London, UK

Institute of Laboratory Animal Resources 1980 *Laboratory Animal Management: Nonhuman Primates.* National Academy Press: Washington, DC

Institute for Laboratory Animal Research 1992 *Recognition and alleviation of pain and distress in laboratory animals.* National Academy Press: Washington, DC
http://books.nap.edu/books/0309042755/html/index.html

International Primatological Society 1993 IPS International guidelines for the acquisition, care and breeding of nonhuman primates, Codes of Practice 1-3. *Primate Report 35*: 3-29
http://pin.primate.wisc.edu/ips/codes.txt

International Primatological Society 2007 *IPS International Guidelines for the Acquisition, Care and Breeding of Nonhuman Primates.* International Primatological Society: Bronx, NY
http://www.internationalprimatologicalsociety.org/docs/IPS_International_Guidelines_for_the_Acquisition_Care_and_Breeding_of_Nonhuman_Primates_Second_Edition_2007.pdf

Jerome CP and Szostak L 1987 Environmental enrichment for adult, female baboons (*Papio anubis*). *Laboratory Animal Science 37*: 508-509

Johns Hopkins University and Health System 2001 Restraint techniques for animals - Non human primates. *Animal Care and Use Training (Web site)*: Accessed 02/11/2008
http://www.jhu.edu/animalcare/training_procedures_restraint.html

Judge PG, Griffaton NS and Fincke AM 2006 Conflict management by hamadryas baboons (Papio hamadryas hamadryas) during crowding: a tension-reduction strategy. *American Journal of Primatology 68*(10): 993-1006

Kaplan JR, Manning P and Zucker E 1980 Reduction of mortality due to fighting in a colony of rhesus monkeys (*Macaca mulatta*). *Laboratory Animal Science 30*: 565-570

Kelley TM and Bramblett CA 1981 Urine collection from vervet monkeys by instrumental conditioning. *American Journal of Primatology 1*: 95-97

Kelley ST and Hall AS 1995 Housing. In: Bennett BT, Abee CR and Henrickson R (eds) *Nonhuman Primates in Biomedical Research* pp. 193-209. Academic Press: New York, NY

Kessel-Davenport AL and Gutierrez T 1994 Training captive chimpanzees for movement in a transport box. *The Newsletter 6*(2): 1-2
http://www.awionline.org/Lab_animals/biblio/jo-6.htm

Kessel AL and Brent L 1995 An activity cage for baboons, Part I. *Contemporary Topics in Laboratory Animal Science 34*(6): 74-79

Kessel AL and Brent L 1996 Space utilization by captive-born baboons (*Papio* sp.) before and after provision of structural enrichment. *Animal Welfare 5*: 37-44
http://www.awionline.org/Lab_animals/biblio/aw5-37.htm

Kessel AL and Brent L 1998 Cage toys reduce abnormal behavior in individually housed pigtail macaques. *Journal of Applied Animal Welfare Science 1*: 227-234

Kessel AL and Brent L 2001 The rehabilitation of captive baboons. *Journal of Medical Primatology 30*: 71-80

Kessel AL, Brent L and Walljasper T 1995 Shredded paper as enrichment for infant chimpanzees. *Laboratory Primate Newsletter 34*(4): 4-6
http://www.brown.edu/Research/Primate/lpn34-4.html#shred

Kessler MJ, London WT, Rawlins RG, Gonzales J, Martines HS and Sanches J 1985 Management of a harem breeding colony of rhesus monkeys to reduce trauma-related morbidity and mortality. *Journal of Medical Primatology 13*: 91-98

Keverne EB, Martensz ND and Tuite B 1989 Beta-endorphin concentrations in cerebrospinal fluid of monkeys are influenced by grooming relationships . *Psychoneuroendocrinology 14*(1-2): 155-161

King JE and Norwood VR 1989 Free-environment rooms as alternative housing for squirrel monkeys. In: Segal EF (ed) *Housing, Care and Psychological Wellbeing of Captive and Laboratory Primates* pp. 102-114. Noyes Publications: Park Ridge, NJ

Kitchen AM and Martin AA 1996 The effects of cage size and complexity on the behaviour of captive common marmosets, *Callithrix jacchus jacchus. Laboratory Animals 30*: 317-326

Knezevich M and Fairbanks L 2004 Tooth blunting as a wound reduction strategy in group living vervet monkeys (*Chlorocebus aethiops*). *American Journal of Primatology 62*(Supplement): 45
http://www.asp.org/asp2004/abstractDisplay.cfm?abstractID=744&confEventID=722

Koban TL, Miyamoto M, Donmoyer G and Hammar A 2005 Effects of positive reinforcement training on cortisol, hematology and cardiovascular parameters in cynomolgus macaques (*Macaca fascicularis*). *American Journal of Primatology 66*(Supplement): 148

Kopecky J and Reinhardt V 1991 Comparing the effectiveness of PVC swings versus PVC perches as environmental enrichment objects for caged female rhesus macaques. *Laboratory Primate Newsletter 30*(2): 5-6
http://www.brown.edu/Research/Primate/lpn30-2.html#vik

Kravic MA and McDonald K 2003 Environmental enrichment of nonhuman primates with PVC pipe constructs. *AALAS [American Association for Laboratory Animal Science] 54th National Meeting Official Program*: 138-139

Lam K, Rupniak NMJ and Iversen SD 1991 Use of a grooming and foraging substrate to reduce cage stereotypies in macaques. *Journal of Medical Primatology 20*: 104-109
http://www.awionline.org/Lab_animals/biblio/jmp20-1.htm

Lambeth SP and Bloomsmith MA 1994 A grass foraging device for captive chimpanzees (*Pan troglodytes*). *Animal Welfare 3*: 13-24
http://www.awionline.org/Lab_animals/biblio/aw3-13.htm

Lambeth SP and Bloomsmith MA 1992 Mirrors as enrichment for captive chimpanzees *(Pan troglodytes). Laboratory Animal Science 42*: 261-266

Lambeth SP, Perlman JE and Schapiro SJ 2000 Positive reinforcement training paired with videotape exposure decreases training time investment for a complicated task in female chimpanzees. *American Journal of Primatology 51*(Supplement): 79-80

LAREF 2007a Chapter 5.1. Pair Formation and Pair-Housing of Monkeys. In: Baumans V, Coke C, Green J, Moreau E, Morton D, Patterson-Kane E, Reinhardt A, Reinhardt V and Van Loo P (eds) *Making Lives Easier for Animals in Research Labs: Discussions by the Laboratory Animal Refinement & Enrichment Forum* pp. 95-101. Animal Welfare Institute: Washington, DC
http://www.awionline.org/pubs/LAREF/soc-hous.html

LAREF 2007b Chapter 5.3.1. Post-Operative Care. In: Baumans V, Coke C, Green J, Moreau E, Morton D, Patterson-Kane E, Reinhardt A, Reinhardt V and Van Loo P (eds) *Making Lives Easier for Animals in Research Labs: Discussions by the Laboratory Animal Refinement & Enrichment Forum* pp. 101-103. Animal Welfare Institute: Washington, DC http://www.awionline.org/pubs/LAREF/buffer-post-op

LAREF 2007c Chapter 7.2. Injection and Blood Collection—How to minimize Stress Reactions. In: Baumans V, Coke C, Green J, Moreau E, Morton D, Patterson-Kane E, Reinhardt A, Reinhardt V and Van Loo P (eds) *Making Lives Easier for Animals in Research Labs: Discussions by the Laboratory Animal Refinement & Enrichment Forum* pp. 117-124. Animal Welfare Institute: Washington, DC http://www.awionline.org/pubs/LAREF/working.html#training

LAREF 2007d Chapter 4.3. Feeding Enrichment. In: Baumans V, Coke C, Green J, Moreau E, Morton D, Patterson-Kane E, Reinhardt A, Reinhardt V and Van Loo P (eds) *Making Lives Easier for Animals in Research Labs: Discussions by the Laboratory Animal Refinement & Enrichment Forum* pp. 52-57. Animal Welfare Institute: Washington, DC http://www.awionline.org/pubs/LAREF/enrich.html#feed-prim

LAREF 2007e Chapter 4.5. Mirrors. In: Baumans V, Coke C, Green J, Moreau E, Morton D, Patterson-Kane E, Reinhardt A, Reinhardt V and Van Loo P (eds) *Making Lives Easier for Animals in Research Labs: Discussions by the Laboratory Animal Refinement & Enrichment Forum* pp. 60-61. Animal Welfare Institute: Washington, DC http://www.awionline.org/pubs/LAREF/enrich.html#mirrors

LAREF 2007f Chapter 4.7. Windows. In: Baumans V, Coke C, Green J, Moreau E, Morton D, Patterson-Kane E, Reinhardt A, Reinhardt V and Van Loo P (eds) *Making Lives Easier for Animals in Research Labs: Discussions by the Laboratory Animal Refinement & Enrichment Forum* pp. 65-66. Animal Welfare Institute: Washington, DC http://www.awionline.org/pubs/LAREF/enrich.html#windows

LAREF 2007g Chapter 8.7. Swimming Pool for Macaques. In: Baumans V, Coke C, Green J, Moreau E, Morton D, Patterson-Kane E, Reinhardt A, Reinhardt V and Van Loo P (eds) *Making Lives Easier for Animals in Research Labs: Discussions by the Laboratory Animal Refinement & Enrichment Forum* p.145. Animal Welfare Institute: Washington, DC http://www.awionline.org/pubs/LAREF/safety.html#swim

Laule GE, Thurston RH, Alford PL and Bloomsmith MA 1996 Training to reliably obtain blood and urine samples from a diabetic chimpanzee (*Pan troglodytes*). *Zoo Biology 15*: 587-591

Lee GH, Thom JP and Crockett CM 2005 Factors predicting compatible grooming-contact pairings in four species of laboratory monkeys. *American Journal of Primatology 66*(Supplement): 83-84

Lehman SM and Lessnau RG 1992 Pickle barrels as enrichment objects for rhesus macaques. *Laboratory Animal Science 42*: 392-397

Leon S and Taylor L 1993 Activity patterns of captive and free-ranging long-tailed macaques (*Macaca fascicularis*). *American Zoo and Aquarium Association (AZA) Regional Conference Proceedings*: 171-178

Lethmate J and Dücker G 1973 Studies on self-recognition in a mirror in orang-utans, chimpanzees, gibbons and various other monkey species [article in German]. *Zeitschrift für Tierpsychologie [Ethology] 33*: 248-269

Levison PK, Fester CB, Nieman WH and Findley JD 1964 A method for training unrestrained primates to receive drug injection. *Journal of the Experimental Analysis of Behavior 7*: 253-254

Lindburg DG 1971 The rhesus monkey in North India: an ecological and behavioral study. In: Rosenblum LA (ed) *Primate Behavior: Developments in Field and Laboratory Research, Volume 2* pp. 1-106. Academic Press: New York, NY

Line SW and Morgan KN 1991 The effects of two novel objects on the behaviour of singly caged adult rhesus macaques. *Laboratory Animal Science 41*: 365-369

Line SW, Morgan KN, Markowitz H, Roberts J and Riddell M 1990a Behavioral responses of female long-tailed macaques (*Macaca fascicularis*) to pair formation. *Laboratory Primate Newsletter 29*(4): 1-5
http://www.brown.edu/Research/Primate/lpn29-4.html#line

Line SW, Clarke AS, Markowitz H and Ellman G 1990b Responses of female rhesus macaques to an environmental enrichment apparatus. *Laboratory Animals 24*: 213-220

Ljungberg T, Westlund K and Rydén L 1997 Ethological studies of well-being in two species of macaques after transition from single-cages to housing in social groups (Web site). *EUPREN/EMRG Workshop http://www.dpz.gwdg.de:80/eupren/eupren.htm*: 20-21
http://www.dpz.gwdg.de:80/eupren/eupren.htm

Logsdon S 1994 Enrichment for woolly monkeys. *The Shape of Enrichment 3*(1): 8

Logsdon S 1995 Use of operant conditioning to assist in the medical management of hypertension in woolly monkeys. *American Zoo and Aquarium Association (AZA) Regional Conference Proceedings*: 96-102

Luttrell L, Acker L, Urben M and Reinhardt V 1994 Training a large troop of rhesus macaques to cooperate during catching: Analysis of the time investment. *Animal Welfare 3*: 135-140
http://www.awionline.org/Lab_animals/biblio/aw5train.htm

Lutz CK and Novak MA 1995 Use of foraging racks and shavings as enrichment tools for groups of rhesus monkeys (*Macaca mulatta*). *Zoo Biology 14*: 463-474

Lutz CK and Farrow RA 1996 Foraging device for singly housed longtailed macaques does not reduce stereotypies. *Contemporary Topics in Laboratory Animal Science 35*(3): 75-78

Lutz C, Tiefenbacher S, Jorgenson MJ, Meyer JS and Novak MA 2000 Techniques for collecting saliva from awake, unrestrained, adult monkeys for cortisol assay. *American Journal of Primatology 52*: 93-99

Lynch R 1998 Successful pair-housing of male macaques (*Macaca fascicularis*). *Laboratory Primate Newsletter 37*(1): 4-5
http://www.brown.edu/Research/Primate/lpn37-1.html#pair

Lynch R and Baker D 2000 Primate Enrichment: A room with a view. *Laboratory Primate Newsletter 39*(1): 12
http://www.brown.edu/Research/Primate/lpn39-1.html#room

Mahoney CJ 1992 Some thoughts on psychological enrichment. *Lab Animal 21*(5): 27,29,32-37

Majolo B, Buchanan-Smith HM and Morris K 2003 Factors affecting the successful pairing of unfamiliar common marmoset(*Callithrix jacchus*) females: Preliminary results. *Animal Welfare 12*: 327-337

Maki S, Alford PL, Bloomsmith MA and Franklin J 1989 Food puzzle device simulating termite fishing for captive chimpanzees (*Pan troglodytes*). *American Journal of Primatology 19(Supplement)*: 71-78

Maninger N, Kim JH and Ruppenthal GC 1998 The presence of visual barriers decreases antagonism in group housed pigtail macaques *(Macaca nemestrina)*. *American Journal of Primatology 45*: 193-194

Markowitz H and Line SW 1989 Primate research models and environmental enrichment. In: Segal EF (ed) *Housing, Care and Psychological Wellbeing of Captive and Laboratory Primates* pp. 202-212. Noyes Publications: Park Ridge, NJ

Mason WA 1960 Socially mediated reduction in emotional responses of young rhesus monkeys. *Journal of Abnormal and Social Psychology 60*: 100-110

McCormack K and Megna NL 2001 The effects of privacy walls on aggression in a captive group of rhesus macaques *(Macaca mulatta). American Journal of Primatology 54* (Supplement): 50-51
http://www.asp.org/asp2001/abstractDisplay.cfm?abstractID=110&confEventID=71

McDermott J and Hauser MD 2007 Nonhuman primates prefer slow tempos but dislike music overall. *Cognition 104*: 654-668

McGrew WC, Brennan JA and Russell J 1986 An artificial 'Gum-tree' for marmosets (*Callithrix j. jacchus*). *Zoo Biology 5*: 45-50

McKinley J, Buchanan-Smith HM, Bassett L and Morris K 2003 Training common marmosets (*Callithrix jacchus*) to cooperate during routine laboratory procedures: Ease of training and time investment. *Journal of Applied Animal Welfare Science 6*: 209-220

McNulty JA, Iskandar E and Kyes RC 2004 Time budgets of long-tailed macaques (*Macaca fascicularis*) on Tinjil Island, Indonesia. *American Journal of Primatology 62*(Supplement): 65-66
http://www.asp.org/asp2004/abstractDisplay.cfm?abstractID=715&confEventID=765

Medical Research Council 2004 *MRC Ethics Guide: Best Practice in the Accomodation and Care of Primates used in Scientific Research.* Medical Research Council: London, UK
http://www.mrc.ac.uk/pdf-primate-best-practice.pdf

Meyerson BJ 1986 Ethology in animal quarters. *Acta Physiologica Scandinavica 554*(Supplement): 24-31

Mitchell DS, Wigodsky HS, Peel HH and McCaffrey TA 1980 Operant conditioning permits voluntary, noninvasive measurement of blood pressure in conscious, unrestrained baboons *(Papio cynocephalus). Behavior Research Methods and Instrumentation 12*: 4l92-498

Molzen EM and French JA 1989 The problem of foraging in captive callitrichid primates: Behavioral time budgets and foraging skills. In: Segal EF (ed) *Housing, Care and Psychological Wellbeing of Captive and Laboratory Primates* pp. 89-101. Noyes Publications: Park Ridge, NJ

Morton DB 1990 Adverse effects in amimals and their relevance to refining scientific procedures. *ATLA [Alternatives to Laboratory Animals] 18*: 29-39

Murchison MA 1994 Primary forage feeder for singly-caged macaques. *Laboratory Primate Newsletter 33*(1): 7-8
http://www.brown.edu/Research/Primate/lpn33-1.html#mark

Murchison MA 1995 Forage feeder box for single animal cages. *Laboratory Primate Newsletter 34*(1): 1-2
http://www.brown.edu/Research/Primate/lpn34-1.html#forage

Murphy C 2008 Socializing weaned macaques in kindergartens. *Laboratory Animal Refinement and Enrichment Forum (electronic discussion group; March 25, 2008)*
http://groups.yahoo.com/group/LAREF/members

Murray L, Hartner M and Clark LP 2002 Enhancing postsurgical recovery of pair-housed nonhuman primates (*M. fascicularis*). *Contemporary Topics in Laboratory Animal Science 41*(4): 112-113

Nadler RD, Herndon JG, Metz B, Ferrer AC and Erwin J 1992 Environmental enrichment by varied feeding strategies for individually caged young chimpanzees. In: Erwin J and Landon JC (eds) *Chimpanzee Conservation and Public Health: Environments for the Future* pp. 137-145. Diagnon/Bioqual: Rockville, MD

Nagel V and Kummer H 1974 Variation in cercopithecoid aggressive behavior. In: Holloway R (ed) *Primate Aggression, Territoriality, and Xenophobia* pp. 159-184. Academic Press: New York, NY

Nahon NS 1968 A device and techniques for the atraumatic handling of the sub-human primate. *Laboratory Animal Care [Laboratory Animal Science] 18*: 486-487

Nakamichi M and Asanuma K 1998 Behavioral effects of perches on group-housed adult female Japanese monkeys. *Perceptual and Motor Skills 87*: 707-714

National Center for the Replacement Refinement Reduction of Animals in Research 2006 *Non-human Primates*. National Center for the Replacement, Refinement and Reduction of Animals in Research: London, UK
http://www.nc3rs.org.uk/category.asp?catID=26

National Health and Medical Research Council [NHMRC] Animal Welfare Committee 1997 *Policy on the Use of Non-Human Primates in Medical Research*. National Health and Medical Research Council: Canberra, Australia
http://www7.health.gov.au/nhmrc/ethics/animal/issues/nonhuman.htm

National Research Council 1985 *Guide for the Care and Use of Laboratory Animals, 6th Edition*. National Institutes of Health: Bethesda, MD

National Research Council 1996 *Guide for the Care and Use of Laboratory Animals, 7th Edition*. National Academy Press: Washington, DC
http://www.nap.edu/books/0309053773/html/

National Research Council 1998 *The Psychological Well-Being of Nonhuman Primates* . National Academy Press: Washington, DC
http://books.nap.edu/books/0309052335/html/index.html

Neveu H and Deputte BL 1996 Influence of availability of perches on the behavioral well-being of captive, group-living mangabeys. *American Journal of Primatology 38*: 175-185

Novak MA and Bayne K 1991 Monkey behavior and laboratory issues. *Laboratory Animal Science 41*: 306-307

O'Neill-Wagner PL 1994 When trying to get your monkeys to behave, try perches. *In Touch 1*(2): 6-8
http://www.awionline.org/Lab_animals/biblio/tou-pegg.htm

O'Neill PL, Lauter AC and Weed JL 1997 Curious response of three monkey species to mirrors. *American Zoo and Aquarium Association (AZA) Regional Conference Proceedings*: 95-101

Ochiai OT and Matsuzawa T 2001 Introduction of two wooden climbing frames as environmental enrichment for captive chimpanzees (*Pan troglodytes*) and its assessment. *Japanese Journal of Animal Psychology 51*(1): 1-9

Öbrink KJ and Rehbinder C 1999 Animal definition: a necessity for the validity of animal experiments? *Laboratory Animals 22*: 121-130

Ott JN 1974 The importance of laboratory lighting as an experimental variable. In: Magalhaes H (ed) *Environmental Variables in Animal Experimentation* pp. 39-57. Bucknell University: Lewisburg, PA

Panneton M, Alleyn S and Kelly N 2001 Chair restraint for squirrel monkeys. *AALAS [American Association for Laboratory Animal Science] 52nd National Meeting Official Program*: 92

Paquette D and Prescott J 1988 Use of novel objects to enhance environments of captive chimpanzees. *Zoo Biology 7*: 15-23

Perlman J, Guhad FA, Lambeth S, Fleming T, Lee D, Martino M and Schapiro S 2001 Using positive reinforcement training techniques to facilitate the assessment of parasites in captive chimpanzees. *American Journal of Primatology 54*(Supplement): 56
http://www.asp.org/asp2001/abstractDisplay.cfm?abstractID=122&confEventID=69

Perlman JE, Bowsher TR, Braccini SN, Kuehl TJ and Schaprio SJ 2003 Using positive reinforcement training techniques to facilitate the collection of semen in chimpanzees (*Pan troglodytes*). *American Journal of Primatology 60*(Supplement): 77-78
http://www.asp.org/asp2003/abstractDisplay.cfm?abstractID=503&confEventID=516

Peterson G, Kelly K and Miller L 1988 Use of an artificial gum-tree feeder for marmosets. *Animal Keepers Forum 15*: 396-401

Phillippi-Falkenstein K and Clarke MR 1992 Procedure for training corral-living rhesus monkeys for fecal and blood-sample collection. *Laboratory Animal Science 42*: 83-85

Platt MM and Thompson RL 1985 Mirror responses in a Japanese macaque troop (Arashiyama West). *Primates 26*: 300-314

Poole TB 1990 Environmental enrichment for marmosets. *Animal Technology 41*: 81-86
http://www.awionline.org/Lab_animals/biblio/at-poole.html

Pranger LA, Maier A, Coleman K, Lambeth SP, Perlman JE, Thiele E, McMillam JL and Schapiro SJ 2006 Venipuncture training using positive reinforcement training techniques: a comparison of chimpanzee and rhesus macaques. *American Journal of Primatology 68*(Supplement): 61-62

Priest GM 1990 The use of operant conditioning in training husbandry behavior with captive exotic animals. *Proceedings of the National American Association of Zoo Keepers Conference 16*: 94-108

Priest GM 1991 Training a diabetic drill (*Mandrillus leucophaeus*) to accept insulin injections and venipuncture. *Laboratory Primate Newsletter 30*(1): 1-4
http://www.brown.edu/Research/Primate/lpn30-1.html#loon

Primate Research Institute 2003 *Guide of the Care and Use of Laboratory Primates (Second Edition)*. Primate Research Institute of Kyoto University: Kyoto, Japan
http://www.pri.kyoto-u.ac.jp/research/en/

Prist P, Pizzutto CS and Hashimoto C 2004 Woven vine balls and baskets as feeding enrichment for howler monkeys. *Shape of Enrichment 14*(2): 1-2
http://www.enrichment.org/articles/142Howlers.pdf

Pruetz JD and Bloomsmith MA 1992 Comparing two manipulable objects as enrichment for captive chimpanzees. *Animal Welfare 1*: 127-137
http://www.awionline.org/Lab_animals/biblio/aw1-12.htm

Public Health Service (PHS) 1996 U.S. Government Principles for the Utilization and Care of Vertebrate Animals Used in Testing, Research, and Training. In: National Research Council *Guide for the Care and Use of Laboratory Animals* pp. 117-118. National Academy Press: Washington, DC

Rawlins J 2005 Stock tanks for yearlong primate enrichment. *Tech Talk 10*(3): 1-2
http://www.aalas.org/pdfUtility.aspx?pdf=TT/10_3.pdf

Reasinger DJ and Rogers JR 2001 Ideas of improving living conditions of non-human primates by improving cage design. *Contemporary Topics in Laboratory Animal Science 40*(4): 89

Reinhardt V 1989 Evaluation of the long-term effectiveness of two environmental enrichment objects for singly caged rhesus macaques. *Lab Animal 18*(6): 31-33
http://www.awionline.org/Lab_animals/biblio/la-eval.htm

Reinhardt V 1990a Time budget of caged rhesus monkeys exposed to a companion, a PVC perch and a piece of wood for an extended time. *American Journal of Primatology 20*: 51-56

Reinhardt V 1990b Social enrichment for laboratory primates: A critical review. *Laboratory Primate Newsletter 29*(3): 7-11
http://www.brown.edu/Research/Primate/lpn29-3.html#rev

Reinhardt V 1990c Avoiding undue stress: Catching individual animals in groups of rhesus monkeys. *Lab Animal 19*(6): 52-53
http://www.awionline.org/Lab_animals/biblio/la-avoid.htm

Reinhardt V 1990d Environmental enrichment program for caged stump-tailed macaques (*Macaca arctoides). Laboratory Primate Newsletter 29*(2): 10-11
http://www.brown.edu/Research/Primate/lpn29-2.html#stump

Reinhardt V 1991a Group formation of previously single-caged adult rhesus macaques for the purpose of environmental enrichment. *Journal of Experimental Animal Science 34*: 110-115
http://www.awionline.org/Lab_animals/biblio/es34-1.html

Reinhardt V 1991b Training adult male rhesus monkeys to actively cooperate during in-homecage venipuncture. *Animal Technology 42*: 11-17
http://www.awionline.org/Lab_animals/biblio/at11.htm

Reinhardt V 1992a Space utilization by captive rhesus macaques. *Animal Technology 43*: 11-17
http://www.awionline.org/Lab_animals/biblio/at.htm

Reinhardt V 1992b Environmental enrichment branches that do not clog drains. *Laboratory Primate Newsletter 31* (2): 8
http://www.brown.edu/Research/Primate/lpn31-2.html#branch

Reinhardt V 1993a Using the mesh ceiling as a food puzzle to encourage foraging behaviour in caged rhesus macaques (*Macaca mulatta*). *Animal Welfare 2*: 165-172
http://www.awionline.org/Lab_animals/biblio/aw3mesh.htm

Reinhardt V 1993b Enticing nonhuman primates to forage for their standard biscuit ration. *Zoo Biology 12*: 307-312
http://www.awionline.org/Lab_animals/biblio/zb12-30.htm

Reinhardt V 1993c Promoting increased foraging behaviour in caged stumptailed macaques. *Folia Primatologica 61*: 47-51

Reinhardt V 1994a Pair-housing rather than single-housing for laboratory rhesus macaques. *Journal of Medical Primatology 23*: 426-431
http://www.awionline.org/Lab_animals/biblio/jmp23.htm

Reinhardt V 1994b Social enrichment for previously single-caged stumptail macaques. *Animal Technology 5*: 37-41
http://www.awionline.org/Lab_animals/biblio/at37.htm

Reinhardt V 1997a The Wisconsin Gnawing Stick. *Animal Welfare Information Center (AWIC) Newsletter 7*(3-4): 11-12
http://www.nal.usda.gov/awic/newsletters/v7n3/7n3reinh.htm

Reinhardt V 1997b Lighting conditions for laboratory monkeys: Are they adequate? *Animal Welfare Information Center (AWIC) Newsletter 8*(2): 3-6
http://www.nal.usda.gov/awic/newsletters/v8n2/8n2reinh.htm

Reinhardt V 1999 Pair-housing overcomes self-biting behavior in macaques. *Laboratory Primate Newsletter 38*(1): 4
http://www.brown.edu/Research/Primate/lpn38-1.html#pair

Reinhardt V 2008 Rhesus macaque kindergarten. *Laboratory Animal Refinement and Enrichment Forum (electronic discussion group)*: March 25, 2008
http://groups.yahoo.com/group/LAREF/members

Reinhardt V and Cowley D 1990 Training stumptailed monkeys to cooperate during in-homecage treatment. *Laboratory Primate Newsletter 29*(4): 9-10
http://www.brown.edu/Research/Primate/lpn29-4.html#vik

Reinhardt V and Reinhardt A 1991 Impact of a privacy panel on the behavior of caged female rhesus monkeys living in pairs. *Journal of Experimental Animal Science 34*: 55-58
http://www.awionline.org/Lab_animals/biblio/es34-5~1.htm

Reinhardt V and Cowley D 1992 In-homecage blood collection from conscious stump-tailed macaques. *Animal Welfare 1*: 249-255
http://www.awionline.org/Lab_animals/biblio/aw1blood.htm

Reinhardt V and Hurwitz S 1993 Evaluation of social enrichment for aged rhesus macaques. *Animal Technology 44*: 53-57
http://www.awionline.org/Lab_animals/biblio/at44.htm

Reinhardt V and Reinhardt A 2000 The lower row monkey cage: An overlooked variable in biomedical research. *Journal of Applied Animal Welfare Science 3*: 141-149
http://www.awionline.org/Lab_animals/biblio/jaaws1.htm

Reinhardt V and Reinhardt A 2002 Introduction. In: Reinhardt V and Reinhardt A (eds.) *Comfortable Quarters for Laboratory Animals, Ninth Edition* ii-iv. Animal Welfare Institute: Washington, DC
http://www.awionline.org/pubs/cq02/intro.html

Reinhardt V, Cowley D, Scheffler J, Vertein R and Wegner F 1990 Cortisol response of female rhesus monkeys to venipuncture in homecage versus venipuncture in restraint apparatus. *Journal of Medical Primatology 19*: 601-606
http://www.awionline.org/Lab_animals/biblio/jmp19.htm

Reinhardt V, Houser WD, Eisele S, Cowley D and Vertein R 1988 Behavior responses of unrelated rhesus monkey females paired for the purpose of environmental enrichment. *American Journal of Primatology 14*: 135-140
http://www.brown.edu/Research/Primate/lpn26-2.html#vik

Reinhardt V, Cowley D and Eisele S 1991 Serum cortisol concentrations of single-housed and isosexually pair-housed adult rhesus macaques. *Journal of Experimental Animal Science 34*: 73-76
http://www.awionline.org/Lab_animals/biblio/es34-7~1.htm

Richmond J 2002 Refinement, reduction, and replacement of animal use for regulatory testing: Future improvements and implementation within the regulatory framework. *ILAR [Institute for Laboratory Animal Research] Journal 43*(Supplement): 63-68
http://dels.nas.edu/ilar_n/ilarjournal/43_supp/v43supRichmond.shtml

Ricker RB, Williams LE, Brady AG, Gibson SV and Abee CR 1995 Environmental enhancement for laboratory-housed squirrel monkeys: Fifteen-year retrospective analysis of procedures. *Contemporary Topics in Laboratory Animal Science 34*(4): 55

Riviello MC 1995 The use of feeding board as an environmental enrichment device for tufted capuchin monkeys *(Cebus apella). Primate Report 42*: 23-24
http://www.awionline.org/Lab_animals/biblio/pr42-2-2.htm

Robbins DQ, Zwick H, Leedy M and Stearns G 1986 Acute restraint device for rhesus monkeys. *Laboratory Animal Science 36*: 68-70

Roberts RL, Roytburd LA and Newman JD 1999 Puzzle feeders and gum feeders as environmental enrichment for common marmosets. *Contemporary Topics in Laboratory Animal Science 38*(5): 27-31

Roberts SJ and Platt ML 2005 Effects of isosexual pair-housing on biomedical implants and study participation in male macaques. *Contemporary Topics in Laboratory Animal Science 44*(5): 13-18
http://www.aalas.org/pdfUtility.aspx?pdf=CT/44_05_02.pdf

Rosenberg DP and Kesel ML 1994 Old-World monkeys. In: Rollin BE and Kesel ML (eds) *The Experimental Animal in Biomedical Research. Volume II, Care, Husbandry, and Well-Being - An Overview by Species* pp. 457-483. CPR Press: Boca Raton, FL

Ross PW and Everitt JI 1988 A nylon ball device for primate environmental enrichment. *Laboratory Animal Science 38*(4): 481-483

Ross SR and Lukas KE 2006 Use of space in a non-naturalistic environment by chimpanzees (*Pan troglodytes*) and lowland gorillas (*Gorilla gorilla gorilla*) . *Applied Animal Behaviour Science 96*: 143-152

Rowell TE 1967 A quantitative comparison of the behaviour of a wild and a caged baboon group. *Animal Behaviour 15*: 499-509

Russell JL, Taglialatela JP and Hopkins WD 2006 The use of positive reinforcement training in chimpanzees (*Pan troglodytes*) for voluntary presentation for IM injections. *American Journal of Primatology 68*(Supplement): 122

Russell, WMS and Burch, RL 1959 *The Principles of Humane Experimental Technique*. Methuen & Co.: London, UK
http://altweb.jhsph.edu/publications/humane_exp/het-toc.htm

Russell WMS 2002 The ill-effects of uncomfortable quarters. In: Reinhardt V and Reinhardt A (eds.) *Comfortable Quarters for Laboratory Animals, Ninth Edition* pp. 1-5. Animal Welfare Institute: Washington, DC
http://www.awionline.org/pubs/cq02/Cq-intro.html

Russow L-M 2002 Ethical implications of the human-animal bond. *ILAR [Institute for Laboratory Animal Research] Journal 43*(1): 33-37
http://dels.nas.edu/ilar_n/ilarjournal/43_1/Implications.shtml

Sackett D, Oswald M and Erwin J 1975 Aggression among captive female pigtail monkeys in all-female and harem groups. *Journal of Biological Psychology 17*: 17-20

Salzen EA 1989 A closed colony of squirrel monkeys for laboratory studies. In: Segal EF (ed) *Housing, Care and Psychological Wellbeing of Captive and Laboratory Primates* pp. 115-134. Noyes Publications: Park Ridge, NJ

Savastano G, Hanson A and McCann C 2003 The development of an operant conditioning training program for New World priamtes at the Bronx Zoo. *Journal of Applied Animal Welfare Science 6*: 247-261

Schapiro SJ and Bushong D 1994 Effects of enrichment on veterinary treatment of laboratory rhesus macaques (*Macaca mulatta*). *Animal Welfare 3*: 25-36
http://www.awionline.org/Lab_animals/biblio/aw3-25.htm

Schapiro SJ and Bloomsmith MA 1995 Behavioral effects of enrichment on singly-housed, yearling rhesus monkeys: An analysis including three enrichment conditions and a control group. *American Journal of Primatology 35*: 89-101

Schapiro SJ, Lee-Parritz DE, Taylor LL, Watson LM, Bloomsmith MA and Petto AJ 1994 Behavioral management of specific pathogen-free rhesus macaques: Group formation, reproduction, and parental competence. *Laboratory Animal Science 44*: 229-234

Schapiro SJ, Perlman JE, Thiele E and Lambeth S 2005 Training nonhuman primates to perform behaviors useful in biomedical research. *Lab Animal 34*(5): 37-42

Schino G, Scucchi S, Maestripieri D and Turillazzi PG 1988 Allogrooming as a tension-reduction mechanism: A behavioral approach. *American Journal of Primatology 16*: 43-50

Schmidt EM, Dold GM and McIntosh JS 1989 A perch for primate squeeze cages. *Laboratory Animal Science 39*: 166-167

Schultz P 2006 I see myself. *AWI [Animal Welfare Institute] Quarterly 55*(3): 6
http://www.awionline.org/quarterly/see_myself.html

Schwenk B 1992 Bungee jumping monkeys. *Animal Keepers' Forum 19*(12): 437-438

Schwindaman D 1991 The 1985 animal welfare act amendments. In: Novak MA and Petto AJ (eds) *Through the Looking Glass. Issues of Psychological Well-being in Captive Nonhuman Primates* pp. 26-32. American Psychological Association: Washington DC

Scientists Center for Animal Welfare 1987 Consensus recommendations on effective Institutional Animal Care and Use Committees. *Laboratory Animal Science 37*: 11-13

Segerson L and Laule GE 1995 Initiating a training program with gorillas at the North Carolina Zoological Park. *American Zoo and Aquarium Association (AZA) Annual Conference Proceedings* : 488-489

Shefferly N, Fritz J and Howell S 1993 Toys as environmental enrichment for captive juvenile chimpanzees *(Pan troglodytes). Laboratory Primate Newsletter 32*(2): 7-9
http://www.brown.edu/Research/Primate/lpn32-2.html#jo

Shideler SE, Savage A, Ortuño AM, Moorman EA and Lasley BL 1994 Monitoring female reproductive function by measurement of fecal estrogen and progesterone metabolites in the white-faced saki *(Pithecia pithecia). American Journal of Primatology 32*: 95-108

Shimoji M, Bowers CL and Crockett CM 1993 Initial response to introduction of a PVC perch by singly caged *Macaca fascicularis. Laboratory Primate Newsletter 32*(4): 8-11
http://www.brown.edu/Research/Primate/lpn32-4.html#pvc

Shively, CA 2001 *Psychological Well-Being of Laboratory Primates at Oregon Regional Primate Research Center.* Willamette Week Online, March 21, 2001: Portland, OR
http://www.wweek.com/html2/shivreport.html

Shutt K, MacLarnon A, Heistermann M and Semple S 2007 Grooming in Barbary macaques: better to give than receive? *Biology Letters 3*: 231-233

Skoumbourdis EK 2008 Pole-and-collar-and-chair training. *Laboratory Animal Refinement & Enrichment Forum (electronic discussion group)*: January 24, 2008
http://groups.yahoo.com/group/LAREF/members

Smith K, St. Claire M, Shaver C and Olexa P 2004 Use of a shredded paper substrate to ameliorate abnormal self-directed behavior of a chimpanzee (*Pan troglodytes*). *American Journal of Primatology 62*(Supplement): 94-95
http://www.asp.org/asp2004/abstractDisplay.cfm?abstractID=702&confEventID=825

Smith TE, McCallister JM, Gordon SJ and Whittikar M 2004 Quantitative data on training new world primates to urinate. *American Journal of Primatology 64*(1): 83-93

Snowdon CT and Savage A 1989 Psychological well-being of captive primates: General considerations and examples from callitrichids. In: Segal EF (ed) *Housing, Care and Psychological Wellbeing of Captive and Laboratory Primates* pp. 75-88. Noyes Publications: Park Ridge, NJ

Snowdon CT, Savage A and McConnell PB 1985 A breeding colony of cotton-top tamarins (*Saguinus oedipus*). *Laboratory Animal Science 35*: 477-480

Spector MR and Bennett BT 1988 The use of naturally occurring manipulanda can reduce the frequency of cage stereotypy in solitary-housed primates. *The Psychological Well-Being of Captive Primates Conference*: 44-45

Spector MR, Kowalczky MA, Fortman JD and Bennett BT 1994 Design and implementation of a primate foraging tray. *Contemporary Topics in Laboratory Animal Science 33*(5): 54-55

Spragg SDS 1940 Morphine addiction in chimpanzees. *Comparative Psychology Monographs 15*: 1-132

Stahl D, Herrmann F and Kaumanns W 2001 Group formation of a captive all-male group of lion-tailed macaques *(Macaca silenus)*. *Primate Report 59*: 93-108
http://www.dpz.gwdg.de/pr/pr59/stahl.pdf

Steen Z 1995 Effects of enriched food acquisition on activity budgets of two tamarin species at Adelaide Zoo. *International Zoo News 42*: 284-298
http://www.awionline.org/Lab_animals/biblio/izn-ste.htm

Stringfield CE and McNary JK 1998 Operant conditioning for diabetic primates to accept insulin injections. *American Association of Zoo Veterinarians (AAZV)/American Association of Wildlife Veterinarians (AAWV) Joint Conference Proceedings*: 396-397

Taylor L and Owens A 2004 Enclosure use by aged squirrel monkeys (*Saimiri sciureus*). *American Journal of Primatology 62*(Supplement): 85
http://www.asp.org/asp2004/abstractDisplay.cfm?abstractID=800&confEventID=808

Taylor WJ, Brown DA, Lucas-Awad J and Laudenslager ML 1997 Response to temporally distributed feeding schedules in a group of bonnet macaques (*Macaca radiata*). *Laboratory Primate Newsletter 36*(3): 1-3
http://www.brown.edu/Research/Primate/lpn36-3.html#taylor

Teas J, Richie T, Taylor H and Southwick C 1980 Population patterns and behavioral ecology of rhesus monkeys *(Macaca mulatta)* in Nepal. In: Lindburgh DG (ed) *The Macaques: Studies in Ecology, Behavior and Evolution* pp. 247-262. Van Nostrand Reinhold: New York, NY

Terazawa N, Nakayama K and Tomonaga M 2002 Japanese monkeys express preferences for high-ceiling rooms over low-ceiling rooms: Utility value of vertical dimension for caged monkeys. *Anthropological Science 110*: 135

Terry RL 1970 Primate grooming as a tension reduction mechanism. *Journal of Psychology 76*: 129-136

Tiefenbacher S, Lee B, Meyer JS and Spealman RD 2003 Noninvasive technique for the repeated sampling of salivary free cortisol in awake, unrestrained squirrel monkeys. *American Journal of Primatology 60*: 69-75

Turkkan JS, Ator NA, Brady JV and Craven KA 1989 Beyond chronic catheterization in laboratory primates. In: Segal EF (ed) *Housing, Care and Psychological Wellbeing of Captive and Laboratory Primates* pp. 305-322. Noyes Publications: Park Ridge, NJ

United States Department of Agriculture 1991 Title 9, CFR (Code of Federal Register), Part 3. Animal Welfare; Standards; Final Rule. *Federal Register 56*(32): 6426-6505
http://www.nal.usda.gov/awic/legislat/awadog.htm

University of Arizona - IACUC Certification Coordinator 2008 Restraint. *IACUC Learning Module - Primates (Web site)*: Accessed 02/11/2008
http://www.iacuc.arizona.edu/training/primate/rest.html

University of Minnesota - Investigators, and Animal Husbandry and Veterinary Staff 2008 Nonhuman Primates. *Restraint and Handling of Animals* (Web site): Accessed 02/11/2008
http://www.ahc.umn.edu/rar/handling.html

Valerio, DA, Miller, RL, Innes, JRM, Courntey, KD, Pallotta, AJ and Guttmacher, RM 1969 *Macaca mulatta. Management of a Laboratory Breeding Colony.* Academic Press: New York, NY

Vertein R and Reinhardt V 1989 Training female rhesus monkeys to cooperate during in-homecage venipuncture. *Laboratory Primate Newsletter 28*(2): 1-3
http://www.brown.edu/Research/Primate/lpn28-2.html#vik

Videan EN, Fritz J, Howell S and Murphy J 2007 Effects of two types and two genres of music on social behavior in captive chimpanzees (Pan troglodytes). *Journal of the American Association for Laboratory Animal Science [Contemporary Topics in Laboratory Animal Science] 46*(1): 66-70

Videan EN, Fritz J, Schwandt ML, Smith HF and Howell S 2005a Controllability in environmental enrichment for captive chimpanzees (*Pan troglodytes*). *Journal of Applied Animal Welfare Science 8*: 117-130

Videan EN, Fritz J, Murphy J, Borman R, Smith HF and Howell S 2005b Training captive chimpanzees to cooperate for an anesthetic injection. *Lab Animal 34*(5): 43-48

Visalberghi E and Anderson JR 1999 Capuchin Monkeys. In: UFAW [Universities Federation for Animal Welfare] (edited by Poole, T. and English, P.) *The UFAW Handbook on the Care and Management of Laboratory Animals Seventh Edition* pp. 601-610. Blackwell Science: Oxford, UK

Watson DSB 1991 A built-in perch for primate squeeze cages. *Laboratory Animal Science 41*: 378-379

Watson LM 2002 A successful program for same- and cross-age pair-housing adult and subadult male *Macaca fascicularis. Laboratory Primate Newsletter 41*(2): 6-9
http://www.brown.edu/Research/Primate/lpn41-2.html#watson

Watson SL, Gray A, Taylor E, Johnson B, Fahm B, McGee A, Bingham W and Banks P 2002 Efficacy of environmental enrichment for garnett's bushbaby (*Otolemur garnettii*). *American Journal of Primatology 57*: 38-39
http://www.asp.org/asp2002/abstractDisplay.cfm?abstractID=409&confEventID=284

Waugh, C 2002 Coconuts as enrichment item for macaques. *Primate Enrichment Forum (electronic discussion group)*: October 24, 2002
http://www.primate.wisc.edu/pin/pef/

Weed JL, Wagner PO, Byrum R, Parrish S, Knezevich M and Powell DA 2003 Treatment of persistent self-injurious behavior in rhesus monkeys through socialization: A preliminary report. *Contemporary Topics in Laboratory Animal Science 42*(5): 21-23

Weick BG, Perkins SE, Burnett DE, Rice TR and Staley EC 1991 Environmental enrichment objects and singly housed rhesus monkeys: Individual preferences and the restoration of novelty. *Contemporary Topics in Laboratory Animal Science 30*(5): 18

Westlund K 2002 Preference of the vertical dimension of cyno pairs living in high cages. *Laboratory Animal Refinement and Enrichment Forum (electronic discussion group)*: November 28, 2002
http://groups.yahoo.com/group/LAREF/members

Whitney, RA, Johnson, DJ and Cole, WC 1973 *Laboratory Primate Handbook*. Academic Press: New York, NY

Wickings EJ and Nieschlag E 1980 Pituitary response to LRH and TRH stimulation and peripheral steroid hormones in conscious and anaesthetized adult male rhesus monkeys *(Macaca mulatta)*. *Acta Endocrinologica 93*: 287-293

Williams LE, Palughi PJ, Cushman A and Gibson SV 1992 Vegetables as dietary enrichment for *Saimiri*. *American Journal of Primatology 27*: 63-64

Winterborn A 2007 Cooperation counts. *AWI [Animal Welfare Institute] Quarterly 56*(3): 16
http://www.awionline.org/pubs/Quarterly/07-56-03/index.htm

Wolfensohn, SE and Lloyd, M 1994 *Handbook of Laboratory Animal Management and Welfare*. Oxford University Press: New York, NY

Wolfle TL 1987 Control of stress using non-drug approaches. *Journal of the American Veterinary Medical Association 191*: 1219-1221

Woodbeck T and Reinhardt V 1991 Perch use by *Macaca mulatta* in relation to cage location. *Laboratory Primate Newsletter 30*(4): 11-12
http://www.brown.edu/Research/Primate/lpn30-4.html#perch

Wrangham RW 1992 Living naturally: Aspects of wild environments relevant to captive chimpanzee management. In: Erwin J and Landon JC (eds) *Chimpanzee Conservation and Public Health: Environments for the Future* pp. 71-81. Diagnon/Bioqual: Rockville, MD

6. PHOTO CREDITS

#1-5,8,13-34,36,38-44,46-56,59,60,64-68,81,83,84,86-97, 108-111,113,114,117,120,121,124-131,139,143: Viktor Reinhardt

#6: Brent Geoghegan

#7: Steven Miles

#9: S. J. Ball (2008)

#10: Marcus Richardson

#11: Kai Perret

#12: Mike and Christine Gartner

#35,106,133,138: Richard Lynch

#37: Valerie Schoof

#45: Carolyn M. Crockett, Washington National Primate Research Center

#57,58,106,142: Bushmitz Mark Moshe

#61: P.K. Levison; reproduced with permission from Levison PK, Fester CB and Findley JD 1964 A method for training unrestrained primates to receive drug injection. Journal of the Experimental Analysis of Behavior 7: 253-254; Copyright 1964 by the Society for the Experimental Analysis of Behavior, Inc.

#62: Rana Bayrakci

#63: Kate Baker

#69-76,144: Bob Dodsworth

#77: Margaret Whittaker

#78: Jaylan Turkkan

#79: Alison Kulick

#80: Andrew Winterborn

#82: Jean McKinley

#85: Mike and Christine Gartner

#98,137: Anonymous

#99: Arnold Chamove; reproduced with permission of the editor of the Laboratory Primate Newsletter

#100: Samantha Bjone, University of New England, Australia

#101: Katie Eckert

#102,103: Kai Perret

#104: James Anderson

#107: Dan and Kerrie Cartwright

#112: Axel Wolff; reproduced with permission of the editor of the Laboratory Primate Newsletter

#115: Eric F Savage

#116,135: Jennifer Green

#118: R. Goburdhun; reproduced with permission of the Institute of Animal Technology; published in Animal Technology and Welfare 5(2), 115-117, 2004.

#119: Ben Basile

#122: Hannah Buchanan-Smith; reproduced with permission of the editor of the Laboratory Primate Newsletter

#123: Martin Deutsch

#132: ElaineVidean

#134: Bio-Serv

#136,140: Evan MacLean

#141: Peggy O'Neill-Wagner